HARRY FER

INVENTOR AND

By the same Author

THE AVALANCHE ENIGMA

Harry Ferguson

INVENTOR AND PIONEER

Colin Fraser

Old Pond

TO LUISA

First published 1972 (John Murray)
Reprinted 1973
First paperback edition 1998

ISBN 0 9533651 2 3

A catalogue record for this book is available from the British Library

Published by
Old Pond Publishing
104 Valley Road, Ipswich IP1 4PA
United Kingdom

Printed and bound in Great Britain by
St Edmundsbury Press, Suffolk

Contents

PART TWO

PART THREE

Illustrations

Introduction

If we are to eat the soil must be tilled and sown. It has been so since civilisations began and it will remain so until science has lifted man from his dependence upon naturally grown food—if it ever does. For many centuries the soil everywhere was scratched open with wooden or iron instruments wielded by hand; and then in some parts of the world peasants began to harness their livestock to the task of dragging the implement through the soil. This could be termed the first technological breakthrough in cultivation methods, and it is one that even today has not been generally adopted in many developing countries. For example, in certain parts of West Africa, the harnessing of oxen or donkeys to pull a plough is an innovation presently being introduced by foreign experts; and in some places they are having to work against superstitions such as if a man is walking or standing behind a bullock when it passes wind it will make him impotent.

Civilisations uninhibited by such considerations adopted the use of animal power on a massive scale and with great effectiveness. In North America there was even a combine harvester built to be pulled by a team of eighty mules. Reapers, rakes, ploughs, cultivators and all manner of other implements to be operated by animal power reached a high level of design and effectiveness, and agriculture became ever more productive as manpower was siphoned from the land to be absorbed in steam-powered industry. Then, in 1885, the invention of the four-stroke internal combustion engine and its subsequent development for transportation brought with it an opening of new vistas for motive and tractive power on the land. As we shall see, under the stimulus of industry's thirst for labour, as well as under the demands of wartime economies, the internal combustion engine was in some countries brought gradually to the farm.

But it was brought merely as a replacement for animal tractive power; and the same implements that had been used with horses, mules and oxen were coupled behind these 'agricultural motors', as tractors were first called. There was no basic change in the concept of farm mechanisation; one form of motive power was being substituted for another, and there the matter ended: until there appeared on the scene an extraordinary man who, between 1917 and 1935, struggled to invent a new concept of farm mechanisation that would integrate tractor and implement in one unit so that once coupled together it would work as a single tool, yet at the same time allow the implement to be changed easily and quickly to form another unit. The successful development of this concept was the second fundamental breakthrough in farm mechanisation, and it is one that has been adopted the world over. Examine all the tractors built by different companies and enterprises today in Russia, Czechoslovakia, Yugoslavia, Germany, Italy, France, Canada, America, Britain, and Sweden and you will find that about 85 per cent of them incorporate the basic invention of that one man, Harry Ferguson, even if their trade name makes no reference to him.

Ferguson was extraordinary not only because of his successful inventions; he was a man of great complexity and with a character of almost unbelievable contradictions. Somehow he succeeded in combining extremes of subtlety, naïvety, charm, rudeness, brashness, modesty, largesse and pettiness; and the switch from any one to the other could be abrupt and unpredictable. At the same time certain of his qualities were unvarying: he had a clarity of vision and unorthodoxy of thought that were astonishing. Inevitably these were interpreted as crankiness by many, along with some of his other characteristics that were indeed cranky: for example his pernicketiness reached a level of mania in many things. He was also possessed of persistence in anything he undertook—persistence that neither time nor setbacks could erode. Despite his small and spare stature, one could never doubt the power of his character. His taut demeanour and quicksilver movements, his professorial and beaky face with its blue eyes peering critically at an imperfect world from behind rimless glasses, gave the impression of an exceptional man.

My interest in Harry Ferguson stems from a period of about five years that I spent as an instructor lecturing and giving courses on farm machinery bearing his name—machinery so excellent that it aroused in me a profound admiration. I never met its inventor: for a biographer not to meet his subject is considered by some to be an advantage, by others a disadvantage. Be this as it may, I was faced with the task of forming an impression of Ferguson as a man by talking to people who had known him and worked with him, and by reading his correspondence. Thus I travelled in England, Ireland, the United States of America, and Canada to search out people and obtain their recollections of fact, and their opinions, about Ferguson. I was fortunate indeed to meet with co-operation everywhere I went; all the surviving people who had played a significant role in Ferguson's life gave of their time to talk to me. I interviewed a total of sixty-two people and recorded more than 160 hours of tape with them.

Inevitably, many conflicting views emerged as I interviewed different people who had known Ferguson. Whom does one believe in such cases? The somewhat arbitrary nature of my task became increasingly apparent to me as I proceeded. Some issues were simply resolved by adopting the consensus view; in others I felt obliged to accept the word of the person whose memory appeared least coloured by time or, in some cases I must admit, self-interest.

Divergent accounts are, I suppose, the plague of any biographer researching the life of someone survived by many people who can give information about him. I sometimes felt, as I listened to tapes over and over again—in an attempt to capture nuances that might help to unravel a mystery—that it would have been easier to deal with documentation only. Reading it might not always be as interesting as face-to-face encounters, but it might provide more objective evidence.

I did have my share of documentation too, however; I was allowed free access to Ferguson's papers and voluminous correspondence. There was shelf after shelf of thick box files at Abbotswood, his home. There were over a hundred and fifty of them, each packed solid with papers, and though I cannot claim to have read every one of them in detail, I did scan them all and read what seemed important. I spent eight months on the research and often felt I should have spent twice as long; but perhaps if I had it would have been even more difficult to reach conclusions on controversial issues, completely inundated as I would have been by minutiae.

I made a point of visiting as many places where Ferguson had spent extended periods as possible, in an attempt to obtain atmosphere. And just occasionally, when I sat quietly in such places as the library that he used as his office at Abbotswood, or stood in the overgrown vegetable garden where he gave his famous demonstration to Henry Ford I, the scenes I had heard described came to life in my imagination.

I have deliberately not quoted detailed sources as part of the text of this book. I believe the foregoing explanation gives sufficient information about source material. It should be noted, however, that whenever quotes in inverted commas appear in the book, the passage comes verbatim from an interview I recorded, or from letters. But I only used such quotes when they were from people whose memory seemed accurate and when their words tallied in style and fact with what, from other knowledge, I would have expected the personalities concerned to say. No one is going to claim that the words are exactly remembered, but they do reflect the mood of the situation, and the manner and character of the speakers.

Inevitably, then, this book is my subjective account of Ferguson and his work, seen largely through the eyes of others. I must extend my heartfelt gratitude to the many, many people who lent me the experience of their eyes and ears. It is unfortunately not possible to thank all who helped me by name, but special mention must be made of Mrs Betty Sheldon, Ferguson's daughter, and her husband Tony. Their desire to help me write an objective book, and the confidence they granted me, were quite invaluable. I hope all the others who helped will be satisfied when I say that their generous co-operation made this book possible. I realise that many of those I interviewed held positions of high responsibility and that time was for them the most precious of commodities. I am particularly grateful to them for the unstinting way that they dedicated many hours to talking patiently with me, often extending more help than I had any right to expect. Finally, my thanks go to Paddy Fraser—no relation but a member of the clan—who typed the manuscript and made constructive suggestions.

PART ONE

1 *A Hard Childhood*

Northern Ireland is blustery and rain swept, with grey houses squat as though designed to stay out of the windflow that has bent the trees and hedgerows permanently to its will. The country has none of the softness that green fields bring to Devon or Eire. Most of the inhabitants are descended from the Scots who migrated to Ireland during the Plantations encouraged by James I, and until the partition of Ireland they were a Protestant minority in a Catholic country. The character of the people has grown out of this background; the already tough nature of the Scots has been further tempered by the atmosphere of adversity in which they have lived, thus creating the Ulster grit and determination of which they are so proud. Certainly, the small country has produced an inordinately large number of famous men; several American presidents had their origins in Ulster, as did many well-known generals, another indication perhaps of the fighting spirit of its people.

The farmers of Northern Ireland have always had to work hard to make a good living, but James Ferguson was better off than most in the latter half of the last century, for he owned his own farm of just over 100 acres—a large holding for Ulster. He lived at a tiny village called Growell, not far from Dromore in County Down, about 16 miles south of Belfast. The Fergusons of Growell could trace their family in Ireland back to the Plantations from Scotland some two hundred years previously, and they were proud of their long standing in the area. It would be incorrect to assume that the family was well off, but neither did it live in the poverty some journalists have claimed. It is a fact, however, that only by unremitting toil were James Ferguson and his wife Mary able to provide reasonably well for their family of eleven children, three girls and eight boys. The fourth of these was born on November 4th, 1884, and though christened Henry George, he was always called Harry.

Lake House, Growell, the whitewashed farmhouse in which

they lived, was not a happy home. James Ferguson, an austere, bearded, wrath-of-God figure, exercised the sternest discipline on his family. He was a Plymouth Brother and an extreme bigot. For the effect on the Scots who settled as a minority group in Ireland has been not only to breed in them an extra toughness, but also to inspire in many of them an implacable loathing for Roman Catholicism. If we consider the hatred and intolerance which today sparks off such deplorable riots and street fighting between the two religious factions, we get an idea of how deep the anti-Catholic feeling must have been when Ireland was one country and the Protestants feared that a Catholic government might be set up in Dublin to rule the island.

Fortunately, the bigotry on the father's side of the Ferguson family was mitigated slightly by the sweet nature of Mary Ferguson and the progressive attitude of her father, Graham Bell. Bell's first wife had died and he had two other daughters by his second. With their father's encouragement, these two girls became the first women doctors in Ireland. Not surprisingly the Ferguson children found it a relief to get away from the oppressive atmosphere emanating from their father; visits to the Bells at Newry came to mean freedom and happiness, elements missing in the hell-fire-and-damnation upbringing meted out at home.

Harry was a spirited child and began to assert himself from an early age. When about four years old, he was playing in the yard when a farmhand wanted to come by with a wheelbarrow but found the tow-haired child in his way.

'Move, Harry! You're in the way,' he said.

'No,' the child said, defiantly holding his ground, 'you're in Harry's way.'

The child also showed a certain mechanical aptitude. There was a large chest of drawers in the house in which each brother and sister had a drawer for their own particular treasures, and each drawer had its own key. The treasures did not amount to much more than the odd marble, rag doll, broken penknife, piece of string or some pebble precious for its shape or colour, but the children were mystified by the regular unlocking of their drawers and disappearance and reappearance elsewhere of their possessions. It was easy to deduce the identity of the culprit; his cunning was less developed than his ingenuity and his own possessions remained miraculously unmolested. One day an elder sister caught Harry in the flagrant act of taking his own drawer right out, and then successively opening the locks of each drawer below and pulling it out so that he could rifle the contents.

The children were expected to help on the farm from a very early age and at that time there was little or no machinery to make the work easier. Hamerton summed up the agricultural scene when he wrote in his *Sylvan Year*: 'Oh, the toil and endurance that are paid for the bread we eat!'

Tractive power on the land, in the years around the turn of the century, was provided almost exclusively by draft animals, and for these there were quite adequate ploughs, harrows, and other tillage equipment. There were also efficient horse-drawn binders and mowers on the market, and some of the barn milling machinery driven by oil engines was well designed and built. But most of this machinery was only within the reach of estate owners and large farmers.

The sole means of mechanised ploughing in the British Isles when Ferguson was a child were enormous steam-driven rigs. These consisted of traction engines, weighing in the region of 10 tons, with a large winch-drum mounted horizontally under the boiler. A pair of traction engines would work together, each standing at opposite ends of a field and winching a plough to and fro between them. The plough was of the balance type, that is to say it was two ploughs in one, like the reversible ploughs of today that have mouldboards which will turn the furrow-slices to the right or left. One set of mould-boards was at work as the plough was winched across the field in one direction, while the other set was cocked up in the air; at the end of the run, the second set was lowered into the ground for the return journey so that the furrows were always turned towards one side of the field. The traction engine drivers signalled to each other by blowing their steam whistles when the plough was at the end of its run and it was ready to be winched in the other direction. Steam enthusiasts still talk wistfully of the days when the steam whistles of the giant ploughing rigs shrilled out over the countryside; but had they been forced to man them as their daily task they might have been less enthusiastic, for there was arduous work involved in keeping the rigs going. They needed fresh supplies of coal and water about every three hours. Most serious, however, was that steam ploughing tackle—which first came into use in about 1850 and continued until well into this century—could only be afforded by wealthy landowners, and the smaller farmers needed far more economical means of mechanised tillage.

On the Continent, and still applying the principle of hauling the plough to and fro across the field, efforts were made to use electric power to drive the winches. It proved even less economical than steam, except on some sugar beet farms in Germany where the beet-processing factory's steam-raising capacity could be used to generate electricity.

The first attempts to provide direct tractive power for British agriculture were made in the years around that of Ferguson's birth. In 1881, J. Braby published a design for a small three-wheeled steam engine, and in 1882 a man called Grimmer produced something similar. Grimmer's engine, small though it was by standards of the time, still weighed two-and-a-half tons. It was an awkward looking device with enormous front driving wheels and a single small rear wheel to steer it; the plough was mounted under its belly and it was said that it would work from two to eight inches deep in the land around Grimmer's native Wisbech. But Braby's and Grimmer's efforts were doomed to failure; their devices were still too expensive, and those very few who could afford mechanised ploughing preferred the steam winching tackles made by such famous firms as Fowler, Foden, Foster, Burrel, and Wallace and Stevens.

All over Europe and North America, therefore, the draft animal was overwhelmingly the main source of farm power. Even the Canadian prairies were cultivated almost entirely by horse-power, though from about 1900 onwards, gangs of ploughs were sometimes hitched behind 10–12 ton traction engines so that as many as sixteen furrows were being ploughed at once. This was a rapid way of covering a large acreage, but at the same time it was harmful to the land: the weight of such heavy machinery compacted the soil and tended to produce a hard stratum, or pan, under the ploughed surface layers. The presence of such a pan prevents proper drainage and root formation.

Tillage with horses was painfully slow: a team of two could

plough about half an acre a day, and even that was at the cost of exhausting work for the ploughman. Gray was right when he wrote of him homeward plodding his 'weary way'. And added to the sheer effort of controlling the plough all day, and manhandling it around on the headlands, was the tedious job of harnessing and unharnessing the team, rubbing them down, and generally tending to their needs. But perhaps the biggest drawback of all to the use of horses was that in a year a pair consumed the produce of five acres, representing a serious reduction in the land available for cash crops, or for produce which could provide milk or meat rather than merely power. This encroachment by horses on the cash-producing acreage was particularly hard on the small farmer.

Harry Ferguson's personal experiences with horses were also unhappy. Right through his life he remembered with bitterness a particular occasion when, as a child, he accompanied his father to market to buy some cattle. On the return journey, his father rode in the trap while Harry drove the cattle. For almost eight miles he walked, and then about half a mile from home, James Ferguson told his son that he could ride in the trap. In this way, the father had only to open the gate of a field and let the cows in before he himself went into the house, but Harry was left with the unharnessing, rubbing down, feeding and watering of the horse before his day was done.

Harry Ferguson deeply resented this sort of treatment, and as a strong-willed boy he had numerous points of difference with his father. The family edict that the Bible was the only permissible reading matter irked him particularly. Both he and his sisters smuggled books into the house and read them under the bedclothes in their quest for wider knowledge and understanding. Perhaps this clandestine reading was the root cause of the eye trouble that afflicted Ferguson all his life, but it certainly helped to make up for the very basic education he received at the two local schools. His attendance at the first of these schools was of short duration: the master caned another boy for what Ferguson considered unjust reasons. He told the master so and had a bitter row with him.

At fourteen he left school and returned to work on his father's farm where he came to hate the toil of farm work. His small stature—the smallest in the family—and light build, were quite unsuited for heavy manual work. A sister remarked that 'he always gave the impression that he considered himself more put upon than the others' when there was farm work to be done.

He also became rebellious over religion in the home. When he was about sixteen he had violent arguments with visiting lay preachers who were staying in the house, for he refused to accept blindly, as was expected of him, the edicts of the family's particular creed. This urge to enquire and discover the truth had, in fact, been with him from a much earlier age and is illustrated by an anecdote that his brother Joe recounted many years later: when Joe was about fourteen and Harry about ten, they were often told to bring the cattle home from the fields. In the dusk of one particular autumn evening, and with mist forming over the low-lying land, they were driving the cattle towards a small bridge over a stream when suddenly a dense cloud of fog swirled wraithlike out from under it. With their background of what amounted virtually to religious and spiritual hysteria, the boys were convinced they were seeing a ghost. Their fright was increased when the cattle stood for a moment as if petrified and then bolted in the other direction. Joe grabbed his smaller brother by the arm and tried to drag him away, but though he was white with fear, Harry insisted on moving forward to investigate, telling Joe to come with him. The elder brother was scared almost witless and ran off. Only then did Harry give in to his fear and leave also.

2 *Apprenticeship*

It was not surprising that the Ferguson boys were anxious to find a life off the farm. The first-born, Joe, had himself apprenticed in 1895 to Combe-Barbour, the Belfast linen spinners, as a maintenance mechanic. He was particularly interested in cars and motorcycles, but the linen companies offered a much safer career than did the embryonic automotive trade at that time. However, it gradually became evident with each passing year that the automotive trade did have a future, and finally in the autumn of 1901, with some minimal financial help from his father, Joe set up a car and cycle repair shop in a tiny premises in Shankhill Road, Belfast.

His brother Harry was still an unwilling farm labourer, and also an unwilling member of the family circle at Lake House. For his rebelliousness over religion was now crystallising into agnosticism. In the last years of his life he wrote a letter describing that process.

'I was brought up to believe, and did believe into my teens, that there was a Hell of torment into which countless millions of people would go and suffer the most awful agonies for all eternity. As I got older, I began to have doubts about this. It seemed a dreadful thought. These millions going to Hell could not have existed if God had not created them. When He created them, being God, He must have known that they would go to Hell. So why create them?

'These were the kind of thoughts that passed through my mind. My conscience forced me to believe that I should investigate all religions possible, in the light of all the discoveries that man had made in science and in other ways. I did so, and I think I could sum up my final conclusions by saying that I do not believe that anybody will be rewarded or punished in whatever the next world may be, for anything they say they *believe*.

'People cannot help their beliefs. We are forced to what we believe by *evidence*. If we are honest we say what we believe. I believed that all these millions could have escaped Hell by saying they believed in the same beliefs as I did. No, what I now believe is that if we are rewarded or punished in another world, it will not be for what we *believe* but for how we have acted.'

Such reasoning, combined with his loathing for farm work, set him at total variance with his father, and he therefore planned his escape. To emigrate to Canada or America, as did so many other Irishmen, seemed the best solution, and in the autumn of 1902 he decided to go.

His plans were made and the ticket virtually bought, when one Sunday afternoon his brother Joe walked into the kitchen of Lake House. He found his brother standing moodily in front of the open fire, obviously reflecting on his forthcoming departure. But Joe had come to forestall that departure, for he needed someone to drive cars for him occasionally, and he therefore offered Harry an apprenticeship with him at the Shankhill Road workshop. The younger brother, who was also passionately interested in cars and engines, at once dropped any idea of going to Canada. That very evening they set out from Lake House in the gathering dusk and walked over the fields to the nearby village of Hillsborough, from where they took the train to Belfast.

How often it seems that a man's life is moving down a certain path and then a chance occurrence deflects it, leading to great achievements and to benefits for humanity. That hasty decision not to go to Canada was such a point for Ferguson. In Belfast he set about learning his trade of mechanic with boundless enthusiasm, and he discovered a particular aptitude for tuning engines. His first success was on a gas engine that was used for driving the lathes and other machinery in the workshop. The wheezy engine produced little power and overheated so severely that Joe Ferguson set his brother the task of carrying buckets of water to cool it. Thoroughly irritated by the heat, the steam, and the smell, and the poor performance of the engine, the apprentice waited until there was no one else in the workshop and then checked the timing of the exhaust valve. He found that it was opening too late, and when Joe Ferguson and one of the mechanics returned, he showed them his discovery with all the delight of a younger brother outshining the elder for the first time. 'I still remember how "gunked" you looked', he wrote to Joe Ferguson over fifty years later.

This incident, more than anything, led him to concentrate on engine tuning, and with uncompromising perfectionism he built up his talent for making engines go better than ever before. Even today a talent for engine tuning is much valued, but in the first years of the century when internal combustion engines were still cantankerous, really gifted engine tuners were about as rare as white flies. Not surprisingly, therefore, J. B. Ferguson and Co.'s business soon began to develop a discriminating clientele, even if Harry Ferguson skidded the first car he ever drove, complete with its owner, through a shop window.

Over a period of several years the repair shop was gradually established as the best in Belfast and it developed a secure financial base, despite a bank's refusal to lend it a few hundred pounds because there was 'no future for the automobile'.

During these years Harry Ferguson attended evening courses at the Belfast Technical College, and the critics who later claimed that he was an 'engineering illiterate who couldn't even read a blueprint' were wide of the mark. It was at the Belfast Technical College that he first met John Lloyd Williams who was to become a friend and partner. Williams was in fact the only intimate friend that he ever had and, more important, retained. For his friendships usually crumbled when differences of opinion arose; it was difficult, if not impossible, to disagree with him and remain his friend. During this period when he was working with his brother, he also became acquainted with T. McGregor Greer, a wealthy landowner with a passion for cars. Ferguson quite frequently went to the McGregor Greer estate of Tullylagen between Cookstown and Dungannon, to the west of Lough Neagh, to repair the various cars that McGregor Greer owned in quick succession. As a mechanic, he slept over the harness room and ate his meals in the kitchen, but when the family got to know him better, their admiration for his obvious technical competence grew into a deep regard for him as a man, and for his qualities of energy and enthusiasm. His connection with the McGregor Greers turned out to be of vital importance.

3 *The Mad Mechanic*

Nowadays, marketing and sales promotion are considered to be a specialised field, almost a discipline. But many years before such techniques of image building were generally recognised, Ferguson was using them; he had an extraordinary gift as a promoter and salesman, an innate understanding of how to set about building up an image for a business. This understanding led him to take part in motor sport as a means of publicising his brother's business, and as early as 1904 he began riding motor-cycles in various events in Ireland. In these ventures he had the advantage of being fearless to an apparently foolhardy degree, and such unpleasant experiences as having a motor-cycle fuel-tank blow up under him left him undeterred. Ferguson was not mad or wild in the conventional sense, however; indeed at first sight he appeared the most sane of young men—fresh, perhaps, but well restrained. He was always neatly dressed, often with a flower or leaf in his buttonhole, and his abundant fair hair was tidily brushed from his high forehead. But below the straight eyebrows lay the clue to his character; his gaze was of an unwavering firmness, and the keenness of those azure eyes, as well as the set of the slightly downturned mouth, betrayed a determination to be defeated by nothing, least of all by his own fear. Hence his calculated physical courage that others considered irresponsible madness. He won races and trials one after the other and each victory was used as a means of publicising J. B. Ferguson and Co. He even persuaded the more cautious Joe to enter a few events, but Joe eventually ran into a yokel on a bicycle and lost all further interest.

One particular event, the Muratti Trophy of 1907, sheds an intriguing light on Ferguson's character: at the end of the 200-mile motor-cycle trial it was found that he and another competitor called Stewart were exactly equal on points, even taking into account the secret controls. The committee suggested that the two riders should each hold the trophy for six months, or alternatively they should ride another 200-mile round to decide the winner. Ferguson refused both solutions, declaring bluntly that he had already won the event and lodging an official protest against Stewart for having tinkered with his motor-cycle while in the Belfast and Limavady controls. Stewart counter-protested against Ferguson, and the committee dismissed both protests. Yet again Ferguson was asked to re-ride the course, but he again refused; the trophy was therefore awarded to Stewart. The interesting aspect of the incident is that it gained a large amount of attention in the press, which doubtless suited Ferguson's ends very well. Who can know whether his customary refusal to compromise might not, on that occasion, have been reinforced by the knowledge of the publicity it would gain?

In 1908, Harry Ferguson became seriously interested in the possibilities of the aviation industry. Accompanied by John Williams, he went to air meetings at Rheims and Blackpool and, with his enthusiasm aroused by what they had seen, decided to build his own aeroplane. A few others had by then been built in Ireland, but none had flown or looked much as though they could. Elsewhere flying machines were making quite good progress, but Britain, and particularly Ireland, was something of an aviation backwater.

Ferguson took some measurements of the aeroplanes at Rheims and Blackpool and once back in Belfast persuaded his brother that to build and fly one would be good for their business. The first engine to be acquired for the monoplane under construction was a Green, and when it arrived Ferguson was so impatient to bench test it that he substituted an old pulley bound with brass wire for the non-existent fly wheel. His impatience might have killed someone, for immediately the engine started this pulley disintegrated, blasting fragments in all directions. Before the monoplane, which was reminiscent of the Blériot, was finished it was fitted instead with an eight-cylinder air-cooled 35 hp J.A.P. engine.

With wings detached and tail resting in the back of a car, the aircraft was towed through the streets of Belfast and out to Hillsborough Park, the estate of Lord Downshire, not far from the Ferguson home at Growell. The first attempts to get it off the ground were beset by troubles, most of them apparently caused by the unsuitability of the propeller (or 'tractor' as it was then frequently called). Indeed, the first propeller fitted, a Beedle, was a strange affair with two scimitar-shaped blades more reminiscent of a design by Leonardo da Vinci than the relatively advanced state of aviation of 1909. Ferguson replaced this with a Cochrane propeller, and though there was an immediate improvement in performance, there were still endless adjustments to be made. Bad weather and propeller troubles limited his attempts at flying to unsatisfactory short hops, until the last day of December 1909. That day was still blustery and bleak, but after wind and rain had prevented him from making any attempt for almost a week, he became determined to make the first flight in Ireland before the year was out. He refused to listen to friends who said he should wait for better conditions.

A reporter from the *Belfast Telegraph* described the scene '. . . The roar of the eight cylinders was like the sound of a Gatling gun in action. The machine was set against the wind, and all force being developed, the splendid pull of the new propeller swept the big aeroplane along as Mr Ferguson advanced the lever. Presently, at the movement of the pedal, the aeroplane rose into the air at a height from nine to twelve feet, amidst the hearty cheers of the onlookers. The poise of the machine was perfect and Mr Ferguson made a splendid flight of 130 yards. Although fierce gusts of wind made the machine wobble a little, twice the navigator steadied her by bringing her head to the wind. Then he brought the machine to earth safely after having accomplished probably the most successful initial flight that has ever been attempted upon an aeroplane.' On that bleak December day, as well as making the first flight in Ireland, Ferguson became the first Briton to build and fly his own aeroplane.

The January 8, 1910, issue of *Flight* magazine reported that Ferguson—who throughout his life suffered from neither false modesty nor a tendency to understatement when making declarations to the press—intended to be the first to fly the Irish Channel.

Correspondence in the columns of *Flight* magazine during the spring of 1910 reveals something of the self-assurance and aggressiveness of Ferguson as a young man. It also reveals the gauche literary style one would expect in view of his scanty education.

'You may be interested to know,' he wrote in a letter to the magazine, 'that out of twelve different tractors (propellers) I have not been able to get one satisfactory. This has kept me back very badly but I have now one on order from Clarke and Co. of Kingston-on-Thames . . . and hope soon to be able to report more satisfactory progress to you. From what my machine has done with poor type propellers, I am quite certain of good results with properly designed ones.'

This provoked a sharp reaction—one can well imagine the English aviation pundits wondering how anyone sitting in the Irish bogs dared to be so bumptious. The next issue of *Flight* printed two sarcastic replies over the signatures 'Propeller' and 'Humble Bee'. 'Propeller' wrote:

'For the benefit of future builders of machines, it is worthy of note that, according to Mr Ferguson, all the above propellers are tarred with one brush, i.e. they are "poor type propellers". In conclusion, may I respectfully ask Mr Ferguson if it would be necessary, in the event of my building a machine, to experiment with a dozen propellers and two pairs of wings as he has?'

'Humble Bee' wrote:

'I see that Mr Ferguson has written about his monoplane that he has advertised so much. . . . The only performance of the machine so far has been a glide which was duly recorded as an historical flight. I have seen the machine and I doubt very much that it is the fault of the propellers that it does not fly.'

Cochrane, the maker of the propeller with which Ferguson had made his first flight, wrote to say he would be 'sorry to suggest that Mr Ferguson's contrariness is the outcome of anything more than pure Irish excitement'.

Ferguson, always ready to join battle, spoken or written, at once replied in *Flight*:

'"Propeller's" letter—As he seems in need of a little advice and guidance I will do all I can to help him.

'Firstly, I beg to inform him that it is always well to read a letter carefully before commenting on it. Did "Propeller" do this?

'Secondly, I would strongly advise him to build not two pairs of wings but ten, because anyone so obviously devoid of common sense could not possibly make a successful machine inside that number.

'. . . Humble Bee shows a lot of sense in hiding his name—that is if he knows as much about my machine or any other as I credit him with knowing. Nothing could give me greater pleasure than to have his opinion on why I have not flown as successfully as a machine could in that time. I am sure it would also interest all your readers, as all admit they have much to learn yet, and therefore, on behalf of them all, I ask for knowledge.'

This letter silenced his critics, at least temporarily.

In April 1910, Harry Ferguson took his monoplane to Massarene Park, County Antrim, but his flights there were not over successful, although he did manage one of a mile. In June he moved to Magilligan Strand, a fine sandy beach in County Derry in the most northern part of the country. There he was more successful and some of his flights were 2½ miles in length at heights of up to 40 feet. These flights at Magilligan Strand were in preparation for the first aviation event in Ireland. The town of Newcastle, about 30 miles south of Belfast, had offered a prize of £100 (then about $500) for the first person making a 3 mile flight there. They therefore organised a 'Grand Aerial Display and Sports Meeting' for July 23rd, 1910. Ferguson, who was the only competitor for the prize, moved his monoplane from Magilligan Strand and set up headquarters in the enormous Slieve Donard Hotel that stands on a hummock overlooking the beach at Newcastle.

While special excursion trains brought the spectators in droves, Ferguson, in spotless white overalls, supervised the assembly of the monoplane that was spread out over the hotel lawn. When it was finally ready the weather conditions were yet again unfavourable, and the topography did not help because Newcastle is 'where the Mountains of Mourne come down to the sea'. The prevailing south-west winds that come gusting down the slopes create much turbulence on the beach.

Finally, at 7.45 in the evening, Ferguson decided to attempt a flight even though it was still gusty. After quite a short take-off run, the aeroplane lurched uncertainly into the air and from a height of about 10 feet nose-dived into the sand, smashing the propeller and a wheel. The Ferguson team was accustomed to such mishaps and at once replaced the damaged parts. Again he tried to get airborne, and exactly the same thing happened. The few spectators who had remained jeered derisively and went home.

The Newcastle authorities realised that the weather had been unfavourable and left their offer open for a month. Ferguson made repeated attempts, for he was not to be put off, but each ended in a crash. He consumed three propellers, three wheels, two wings and yards of stays. It is remarkable that he did not break his own neck, and it is hardly surprising that he was frequently referred to, in the press, as the 'plucky Irish aviator'. This at least was better than the 'Mad Mechanic of Belfast' which had been one of his earlier press titles. But not everyone was tolerant of his failures; a number of people wrote him sarcastic anonymous cards. One was addressed to H. Ferguson Esq., Aviator on the Ground, and another to Lord Swank Ferguson, Bluff Aviator. The first said: 'Who said he could fly? Who can't fly? Who shouldn't have said he could fly? Who ought to sell his aeroplane and buy a kite? Who are we sick of?—Harry Ferguson.'

The second attached a witty article called 'Postponed', the last part of which ran: 'He has not done it yet. On the first day, the wind was too high. On the second, the barometer was too low. On the third, he could not find his flying boots. On the fourth, conditions were unfavourable, and so on. But today, everything seemed favourable and it was reported that the Wizard of the Skies was getting up.

'Excitement reached fever pitch. A huge crowd collected in the paddock. An enthusiast threw his cap into the air.

'With a grave face, the aviator's secretary stepped forward.

'"Gentlemen," he announced, "I regret to state that owing to a member of the crowd having just thrown his cap up, and disturbed the air currents, there will be no flight today."'

The person who sent this clipping to Ferguson had written on the card to which it was attached. 'You can't fly, only gull the Newcastle U.D.C.'

Finally, on August 8, he managed to get airborne from Dundrum Bay about 2 miles north of Newcastle and at heights varying from 50 to 150 feet flew along the front. As he came soaring over the Slieve Donard Hotel the spectators went wild with excitement. When he had covered 3 miles, he landed and was carried back shoulder high to a dinner and cheque-awarding ceremony.

Ferguson himself wrote later, 'I am sure I made at least 500 attempts to win the £100 before actually doing so. Sometimes I would get half a mile or a quarter and sometimes only a few hundred yards. Then I would get into a swirl of air, or down-current, or air pocket as they are called, and come crashing down. There was not a piece of the machine that at some time or other did not break at Newcastle. The aero sometimes turned a half somersault after landing, or rather falling, and had it not been for the design of the machine I would assuredly have been killed on many different occasions.'

Following the successful flight at Newcastle he made many others, even carrying a passenger from time to time at Magilligan Strand. Then in October of the same year he had barely taken off when the aircraft was hit by turbulence. Out of control, it pulled up into a steep climb, stalled, dropped a wing and spun into the sand with tremendous impact. Ferguson was knocked unconscious and badly cut and bruised. The aircraft was a total wreck and for the moment his flying was over.

It is worth speculating whether Ferguson had hoped to make an opening into the embryo aircraft industry. He showed great tenacity and courage in flying as well as he did, and even if publicity and promotion had been one of the aims, it would have been out of character for him not to have some other objective prior to embarking on such an enterprise. The most likely solution is that indeed he did hope to enter the aviation industry but that he was canny enough to realise, from his own experience, that the outlay in capital and research before a marketable product could be built would not warrant the effort or returns. If so, it was a long-sighted view of aviation which, compared with the motor industry, has proved itself so complex and unrewarding.

One interesting aspect is that it was probably pilot ineptitude more than design faults that caused him so much trouble. Indeed, his aircraft was highly praised by many aviation experts and journals for its excellent, and in many ways original, design and construction; and his wizardry with engines had transformed the J.A.P. unit from an oil spitting, hot running, temperamental beast giving only 260 pounds pull into one more reliable and better mannered giving 320 pounds pull with the same propeller. The rebuilt aircraft was finally crashed by John Williams in 1913 and only the engine and seat were salvaged. Today they are in a newly created aviation museum in Dublin and at the time of writing it is hoped to build a replica of the Ferguson monoplane.

4 *Independence and Marriage*

Motor-cycle exploits and flights, and perhaps even more so the crashes, made Ferguson into something of a public figure in Ireland. Indeed as a personality he far outshone his employer and older brother Joe, and inevitably this gave rise to resentment. But far more irritating to Joe were the pieces of aeroplane taking up space in his workshop, and the amount of time and money that was being spent on flying. And so relations between the brothers slowly soured: this was in fact predictable, for anyone less suited to the role of complacent employee than Harry Ferguson can hardly be imagined; he had all the wildness of a youth bursting with ideas, and he needed independence to be able to express himself. And in truth, his arrangement with his brother Joe had been doomed from the start, for no two Ferguson brothers or sisters ever agreed on anything for very long, if they agreed at all. Each was too similar to the other, too opinionated and unwilling to compromise. Most of them emigrated to America where they took up the most diverse trades. One of them ran a riding school, and another sold radio parts. Yet another, who was a brilliant chess and draughts player, had a booth at Coney Island and by carrying on ten or twelve games at once with members of the public, and betting considerable sums on his victories, made a good living until he began drinking.

In the last years of Harry Ferguson's life, an old employee asked him what had happened to his brothers Ted, Hughie, Norman and Billie—'All dead! All dead!' said Ferguson so abruptly that no further conversation on the subject was possible. This reaction was certainly in part caused by his unwillingness to talk of the more menial status of his brothers when he had himself reached a pinnacle of success; but even more fundamental was the lack of that strengthened link of affection that often grows in families when a tyrannical father creates a bond of solidarity and love among his children.

In 1911, partly because of the strained relations with his brother,

Harry Ferguson decided to establish his own motor business. He was encouraged by McGregor Greer who offered to finance him; and John Williams and some others also provided capital. At its inception the company was called May Street Motors because Joe Ferguson would not allow Harry to trade under the family name. About a year later, however, the company name was changed to Harry Ferguson Ltd.

May Street Motors was immediately successful, partly because Ferguson was so well known; and even if a few people were irritated by his cocky self-assurance, they none the less admired his abilities. He obtained agencies for several cars, among them Vauxhall and Darracq, and he made the company into an extension of his own personality. His fanatical attention to detail permeated every aspect of his business. One of his early employees said:

'Ferguson was a terribly fussy and pernickety man. Cleanliness was a way of life to him.'

Every car brought in for repair was washed before being returned to the customer, and the inside was cleaned out as well. Vehicles in the garage had to be lined up precisely on a chalk mark and the whole premises was kept scrupulously clean. Every mechanic had to wear clean overalls, and would be reprimanded for not changing them during the day if they got dirty.

Orderliness, cleanliness, and punctuality, were qualities that Ferguson insisted on to such a degree that some people considered him a crank; but others believed that his regard for such matters was based on a philosophy that he put into practice from his earliest days in business. His attitude to apprentices probably sums up this philosophy best.

When a youngster came to him for a job he would make him understand that there were certain conditions attached to his employment. Ferguson would give the boy opportunities to learn, and teach him all he himself knew. But in return, he expected loyalty, and that the employee would do everything that he already knew how to do and which required no brains or training. This included punctuality; if working hours began at 8 a.m., the employee must be there at that time. If Ferguson wanted him to be clean and tidy, he must be so, for he was a representative of the Ferguson company and as such was representing Ferguson himself. If the man drilled a $\frac{7}{16}$ inch hole instead of a half-inch hole, that might be due to stupidity and was therefore excusable; but it required no brains to be punctual and tidy. And if a workman left his tools lying around and the workshop was disorderly, a lot of valuable time would be lost looking for things. In sum, there are so many real problems in need of solution that it is inexcusable to add to them, or make it more difficult to solve them, by failing to pay attention to details which require no brains, but merely call for discipline and loyalty.

It was natural that new employees either observed such standards and entered into the spirit of working with such an exigent man in return for the prestige it brought, or they were out within weeks. And there was considerable prestige in working for him because his business prospered. The service he offered was second to none, and his gift for marketing and salesmanship ensured him a major share of the new car market. Indeed, Charlie Sorenson, Henry Ford's close adviser for forty years, once said that 'Harry Ferguson could sell you the birds off the trees'.

In 1913, Ferguson laid another brick in the foundation on which he later built his success—he married, and some might even say that his wife was the cornerstone of his success. Maureen Watson was the daughter of the owner of the grocery shop at Dromore, a village close to the Ferguson home, and since the Watsons were also Plymouth Brethren the families were friendly.

The children, of which the Watsons had ten—to the Fergusons' eleven—had grown up together. Maureen was beautiful with blonde hair; she was gentle and serene with a kind nature. She was so attractive that quite unwittingly she often stole the interest of her sisters' boy friends. Ferguson was still working with his brother Joe when he first began to take Maureen out and to think seriously about marrying her. It happened that Joe was also in love with her and to see his brother's feelings reciprocated, while his were not, added to the friction between them. John Williams too, when asked many years later why he had never married, replied: 'Because there wasn't another Maureen.'

The Watson parents disapproved strongly of Ferguson. Rather than being impressed by his exploits in the automotive and aviation spheres, they merely considered him a madcap. That he was also an agnostic and would declare so openly, backing up his views with well-reasoned arguments to anyone who cared to tilt at him, was the final point that made him totally undesirable as a friend of their daughter.

Harry Ferguson had only one activity at that time of which the Watsons could approve—his support of the Ulster Loyalists and their anti-Catholicism. Just before World War I the British government seemed on the verge of granting Ireland's demand for Home Rule under a Dublin, and therefore mainly Catholic, parliament. This provoked a crisis and a movement by the Protestants to keep Ireland, or at least Ulster, attached to Britain. The movement was led by Sir Edward (later Lord) Carson, a strong-willed and strong-faced lawyer and politician, who persuaded Ulstermen to take the law into their own hands should it be necessary to do so. Winston Churchill dismissed Carson's brilliant oratory as 'frothings' and totally miscalculated the mood of Ulster when he said that in the end 'civil war would evaporate in uncivil words'. For nearly half a million Ulster people signed, some in their own blood, the famous Ulster League and Covenant which declared their intention of using 'all means that may be found necessary', to overthrow attempts to establish Home Rule in Ireland.

The 'all means' signified that they would fight to the end. The manpower to do so was not lacking, and the Ulster Volunteer Force (U.F.V.), established in 1912, was soon 100,000 strong. Arms were, however, harder to come by, and a continuous traffic in guns and ammunition began; one stout-hearted Ulster-woman from Dungannon came down the gangway of the ship from Liverpool looking as though she were 'within a week of her time—and divil a ha'porth the matter with her, only cartridges'.

The shortage of arms for the Ulster Volunteer Force was solved finally in 1914. A Belfast businessman went to Hamburg and bought 30,000 rifles and three million rounds of ammunition, at a cost of over £70,000. Their shipment to Ulster became a month-long hide and seek chase with the British navy while the press reported wildly on a 'mystery Ulster arms-ship'. In fact there were several vessels involved because the arms were transhipped repeatedly during their meanderings round the British Isles before being brought alongside in Larne at dead of night aboard the *Mountjoy II* (named after the famous ship that broke the siege of Londonderry in 1690).

Ferguson became an active supporter of the U.F.V. by gun running on its behalf. During the unloading of the *Mountjoy II*, McGregor Greer's two daughters were among the lookouts stationed on hill tops to give warning of government forces, while their father and Ferguson were ferrying arms around the countryside in their cars and arranging for them to be hidden.

Despite this anti-Catholic stance, Maureen Watson's parents did their best to dissuade her from seeing Ferguson and she was forced to steal out of the house with one of her sisters in order to meet him secretly.

Maureen showed the most remarkable devotion to Ferguson. She once jumped into the sea at Newcastle after he had told her to do so. The fact that makes this episode remarkable is that she could not swim, but when he told her that to jump into deep water was the only way to learn, she dutifully did so. On the other hand, Ferguson claimed to have given up flying in order to assuage Maureen's fears for his safety. (These fears were in no way helped by one journalist who, on seeing a pennant for the aircraft on which she had embroidered 'Good luck, Harry', acidly asked if she did not mean 'Good-bye, Harry!') The reason for having given up flying seems dubious because by 1912 he was racing cars —an activity almost as dangerous as flying. His company had the agency for Vauxhalls which were fast for that period. He had so much success with the marque in local Irish trials that Vauxhalls offered him a 3 litre works car for the Coupe de l'Automobile that was to be run in conjunction with the Dieppe Grand Prix of 1912.

During practice for the event, Ferguson—who was described by one motoring historian as 'terrifically enthusiastic'—considered that there was a defect in the steering on his car and therefore asked another driver to accompany him for a demonstration. At close to the car's maximum speed of 90 m.p.h., Ferguson stood on the brake pedal; the car skidded and somersaulted twice, finishing upsidedown in a ditch. Miraculously neither driver was hurt, though the car was taken over by another for the race. It was not clear whether this was because Vauxhalls were tired of Ferguson or because he was tired of high-speed somersaults. The first seems more likely.

Maureen Watson and Harry Ferguson decided to marry in 1913 and the wedding was boycotted by both families, though the Watsons relented sufficiently to allow Maureen's sister Florrie to attend. The couple were married in a registry office at Newry by one of Ferguson's maternal uncles, a magistrate and one of the Bell family with whom Ferguson had always got on well. For the first months after their marriage Ferguson refused to allow his wife to wear a wedding ring, for all such ornaments were symbols of slavery, he said. Maureen appreciated these fine sentiments but pointed out that convention required married women to wear a ring and that she felt ill-at-ease without one. Ferguson remained unmoved and Maureen finally bought her own. As it turned out, the gentle, kindly Maureen was to be the hidden force behind her husband for the rest of his life. She was very ambitious on his behalf and did everything within her power to help and encourage him, but always discreetly. Like many mercurial people his spirits tended to be either high or low with little in between. After a setback it was Maureen's gentle influence and support that raised him again to his normal ebullient state.

The couple remained devoted to each other for the forty-seven years of their marriage and this devotion touched all who saw it. On one occasion in his latter years, Ferguson offered a cigar to a journalist who was interviewing him and then quoted the whole of Kipling's 'The Betrothed', the poem in which the narrator is faced with the choice of Maggie, his fiancée, or of his beloved cigars. (Ferguson was very fond of cigars—the only thing he ever smoked and even those he smoked sparingly—and he also loved to quote Kipling at length.) At the end of his recitation of the poem, in which the narrator finally decides that cigars are preferable to women because when you have done with them you can throw them out of the window, Ferguson said: 'There is a lot of truth in it, but I don't believe it at all. A good woman is better than anything on this earth.'

He had more reason to make a remark of that sort than most men, for his wife was a truly remarkable person. No one could claim that he was an easy person to live with, yet somehow his wife succeeded, through her devotion and admiration, in identifying herself completely with him. Whatever he wanted became her own sincere desire. The people that he either liked or disliked, she liked or disliked. She never spoke of 'Harry and I' in any way other than in the context of an accordant 'we': 'Even when Harry is wrong, he's right,' she once remarked to a sister.

It would be easy to assume from this that she completely subjugated her own character to that of her husband, but it was not so. For she retained her own character despite her identification with his wants and needs. She had a better education than her husband and she taught him a great deal in the early years of their marriage. She advised, supported and helped him in all his endeavours and she was the most charming and considerate of hostesses. He himself acknowledged his debt to her when he wrote in his will: 'Any success I have had in business during my life I attribute in fair proportion to the loving and solicitous co-operation of my wife and to her loyal companionship, partnership and unvarying devotion in all things that concerned us.'

5 *The Inventive Streak*

Harry Ferguson's career as an inventor began in 1913 when, with his motor business running well, he turned much of his attention to the improvement of carburetters. He first began to invent because his fetish for accuracy and efficiency was outraged by the crudeness of many of the mechanical devices with which he dealt. Carburettors had always held a great interest for him and, as already mentioned, he was adept at tuning them. However, carburettors of the era were rough and ready devices by today's standards and Ferguson invented two basic improvements for which he was granted patents. These inventions evoked a great deal of favourable comment in the technical press, but one reader of the *Autocar*, who signed himself 'A.C.', wrote a number of letters, published in the columns of the journal, in which he queried the claims made for them. Ferguson replied three or four times but finally lost patience: 'If instead of corresponding about it he makes some simple tests, he will find I am quite right,' he wrote as a final rejoinder.

Ferguson's inventiveness was not of the sort that took him to the workbench to translate his ideas into metal: throughout his life he thought in terms of overall principles, of concepts, and he therefore needed a designer, someone who could grasp his idea and work out the best way of applying it. In other words, he wanted someone to do the tinkering until the practical details had been settled. In this respect, he was extremely lucky, for he had had in his employ, since the inception of May Street Motors, a truly brilliant natural engineer and designer called William, or Willie, Sands.

Sands was only about twenty when he joined Ferguson, after serving an apprenticeship in the maintenance of machinery for the linen industry. But his mechanical interests were very much wider, and in particular he was an enthusiastic motor-cyclist. Long before he began to work for Ferguson, he had seen him competing in motor-cycle trials; and he had also gone to Newcastle to watch some of the flying attempts. When Ferguson established May Street Motors in 1911, Willie Sands was quick to apply for a job. From then until the 1950s, Willie Sands, a serious, sad-eyed man, played a major role in Ferguson's life; and this role was often a dramatic one. Without in any way belittling the genius of Harry Ferguson, it must be quite plainly stated that Willie Sands's contribution to his achievements was fundamental.

Indirectly, it was World War I that set Harry Ferguson along the path that was to bring him fame; for the war brought some basic changes of attitude in British agriculture, and these attracted him back to the farming world he had left when he went to join his brother as an apprentice. During those years of World War I, Britain found herself threatened, as never before in her history, by the prospect of grave food shortages. Owing to the country's traditional dependence on imported foodstuffs, the Germans had an excellent opportunity of trying to starve her into submission by breaking the lifeline of ships bearing food to her shores. And the Germans set about the task with ruthless energy: U-boats prowling the Atlantic began to sink alarming numbers of the ships. It therefore became imperative as never before that Britain should produce more of her own food.

British agriculture, however, was finding it difficult even to maintain its production, let alone increase it, for men were being called away in their hundreds of thousands to the holocaust on the Continent. And horses, too, were needed. The obvious solution lay in agricultural mechanisation, but British farmers generally were loth to view tractors with anything but distaste and the complete conviction that nothing would ever displace the horse.

This attitude had prevailed despite the existence since about 1902 of some crude but really quite effective tractors—'agricultural motors' or 'agricultural self-moving engines' running on 'Texas liquid fuel', in the jargon of the era. Without doubt the best of these and the first practical British farm tractor was the Ivel, designed towards the end of 1901 by Dan Albone of Biggleswade. Albone was an exuberant, energetic man who built bicycles and motor-cycles at his Biggleswade workshop before designing the tractor, which he introduced to the public in August 1902.

The Ivel in its original form was fitted with a 10 hp engine driving through two large rear wheels; steering was by a single front wheel. The whole tractor only weighed 1¼ tons, a remarkable lightweight for that era of gigantic steam locomotives. As a simple and unsophisticated source of tractive power, the Ivel was neatly efficient, and it created a considerable stir when first displayed. Would-be backers offered Dan Albone more capital than he needed to produce his tractor, and it was quickly put on sale at a price of £300 ($1500).

During subsequent years, the Ivel was improved by giving it more power, and it won medal after medal at agricultural shows all over Britain. It could plough, spread manure, pull two binders or two mowers at once, and carry out the innumerable hauling jobs on the farm. At one demonstration, it even pulled a 4½ ton threshing machine over soft ground. In 1906, Albone, the 'human dynamo' as he had been nicknamed, died of a stroke at the age of forty-six, but his co-directors continued the business.

The Ivel tractor had already been exported to twenty-five countries at the time Albone died, and a little later it went into production in North America. But the British farming community remained obstinate in its refusal to adopt the Ivel, or any of the other tractors such as the Saunderson, the Simplex, or the Scott which followed it and which were fairly efficient. This conservatism among farmers endured despite continuous press exhortations: one paper said that every farmer should have an Ivel, and there were repeated reports of the lower cost of farm operations using tractors, particularly for ploughing. In June 1905, for example, the *Implement and Machinery Review* published figures to show that the cost of ploughing an acre was 17 shillings with steam, 14s. 3d. with horses, and 8s. 3d. with an Ivel.

One editorial exhorted members of the machinery trades to 'take time by the forelock and make hay with the agricultural motor while the sun of fewness and newness lights upon it'.

However, there was no cause to fear the 'fewness and newness' wearing off—quite the contrary: the impregnable and lasting attitudes of the farming community were those that emerged at a meeting in Edinburgh in 1906 when Professor John Scott, the designer of the Scott tractor, gave a lecture on tractors in general. He spoke of the saving in cost that they could bring, of the extra crops that could be grown, and of the great convenience of tractors. His audience was unimpressed and voiced its opinions firmly during the discussion following the lecture. Some speakers stated flatly that motors could never replace horses; others said that tractors smelt bad and raised the dust; and still others wanted to know how—if Scott's claim that tractors would allow more food to be grown were true—they would sell their extra produce.

This resistance towards the tractor prevailed right up to, and into, the years of World War I. The sales figures for the Ivel tractor between 1902 and the early 1920s, when the company succumbed to the depression, give a good picture of the lack of progress made by farm mechanisation in those years: only about 1500 Ivels were built, and many of these were exported. It is only fair to point out, however, that the prices of tractors at the time (£300–400 or $1500–2000) put them beyond the reach of small farmers. None the less, even the farmers with large holdings of 300 acres and over stuck to their horses, against all the evidence in favour of tractors, crude as these machines then were by later standards.

In North America, farm mechanisation was further advanced in the years just before the 1914–18 War: they had even built giant steam-powered combine harvesters as early as 1904 (history does not relate how many fields of grain they set alight). By 1910, steam traction engines, pulling their large gangs of ploughs and compacting the soil, were still being widely used; but gasoline engines were also becoming increasingly popular. International Harvester, who had for years been making enormously heavy traction engines, put up a £200,000 production plant for gasoline-engined tractors in the years 1910–11. Few, if any, of these American tractors were imported into Britain in the years leading up to World War I, but then the hour of crisis struck for British agriculture. As one editorial in the *Implement and Machinery Review* said:

'It is no longer a question of men *versus* machinery, for men are not to be had for farm work: and the same stringency applies to horses. A large farmer in the South of England writing to us intimates that he recently had to sell a dozen horses for the want of men to attend them, and for working his remaining 16 horses, he has only 3 men available.'

Yet, at the same time, the authorities were calling for more production from British agriculture to make good the food being sent to the bottom of the Atlantic by German torpedoes. Finally, the farming community was forced to accept that mechanisation was necessary, and the importation of tractors from North America began.

The first effect of the war on Ferguson's business was that John Williams left to join the Royal Flying Corps. (He was subsequently shot down over France and received injuries which left him without the use of one arm.) Secondly, Ferguson began to sell farm tractors; the company took on the agency for an American tractor called the Overtime, a strange looking machine with engine, radiator and fuel tank all mounted independently on a main frame. It weighed 2¼ tons and developed 24 hp. The *Implement and Machinery Review* described it as a 'compact little motor', which is indicative of the criteria of the times.

With his gift for salesmanship, Ferguson realised that the best way to promote the tractor was by giving public demonstrations to the farming community, and he duly laid on many. But in Willie Sands's words:

'Most of the farmers came to laugh at us. They were used to the beautiful ploughing you could do with horses and a one- or two-furrow plough, but they never bothered to think how slow and expensive it was. We used a three-furrow Cockshutt plough behind the Overtime. It only needed one furrow to run slightly shallow or slightly deep, so that the furrow slice either fell flat or stood too much on edge, and a jeer of derision went up from the farmers. I don't know what would have happened if we had ever got the tractor bogged.' And it would not have been difficult to get it bogged in Ireland.

The whole matter of beautiful ploughing was in fact at the root of the British farmer's reluctance to accept tractors. Certainly, with horses it was possible to do a perfect job—not a blade of grass or piece of straw visible anywhere, each furrow cocked at exactly the same angle and at exactly the same height—in fact the whole field left as a pattern of straight lines and perfect symmetry that delighted the eye. The art of good ploughing was a matter of professional pride, and woe betide the ploughman whose furrows were not perfect: the slightest curve was, and for that matter still is in many parts of Britain, an excuse for some scornful and probably colourful comment. (In Devon, for example, where I first learned to plough and quite often sinned against symmetry, I was told that my furrows were like the mark left by a 'boar pig pissin' in the snow'. For the benefit of non-agricultural readers it should perhaps be explained that boars often urinate while walking.)

Even today's tractors and ploughs cannot do as aesthetically pleasing a job as a horse team, but ploughmen have come to temper their desire for artistic accomplishment with the need to do the job quickly and efficiently, for the straightness of the furrow makes no difference to the quality of the crop that will grow. But in the early years of World War I, farmers were still not prepared to give up beautiful furrows for the greater speed and efficiency of tractor ploughing. Only when the U-boat offensive reached its peak in 1916–17 and the work of the Committee on the Home Production of Food, under the Right Honourable Viscount Milner, took on great urgency, was it *generally* recognised that tractors were a key to the nation's survival of the crisis. And it was a grave crisis, for it is said that at one point Britain had available only two weeks' supply of food.

More and more grassland had to be broken up for food production, yet British farms had lost 350,000 members of their labour force by 1916–17. The situation was desperate, and Ireland in particular was called upon to make a special effort towards turning over pasture land to food production. The government asked that Ireland plough an additional half million acres in 1917. Their target was more than reached, for from a total of 2,400,328 acres ploughed in 1916, 3,037,730 acres were broken in 1917. England and Scotland combined only ploughed an additional 350,000 acres that year. Indeed, during the last year of the war, England was receiving more agricultural produce from Ireland than from any other country.

Ferguson and Sands were at least in part responsible for the ploughing success in Ireland. By 1917, as a result of their demonstrations with the Overtime, they had gained a reputation as adept tractor ploughmen—a rarity in those days. The Irish Board of Agriculture therefore asked Ferguson to improve the efficiency of Ireland's tractors during the spring ploughing season of 1917, in order to ensure more cultivated acres off which to feed the people of Britain.

Ferguson took up his duties on March 19th, 1917, and he and Sands set off in a government-supplied car—a large one modified so that they could sleep in it if necessary. Their first start, with Ferguson immaculate but businesslike in trench-coat and highly polished knee boots, was at 5 a.m. on a frosty morning. The uncivilised hour shook Sands, and he was relieved when on subsequent mornings it was at least daylight before they began work. They visited tractor after tractor all over Ireland in the following weeks, and they also laid on innumerable demonstrations. Most of the tractors and ploughs they were dealing with were imported from America; and all of them were cumbersome and extremely complicated to use. The Bates Steel Mule, for example, was the most extraordinary machine: it was a monstrous affair with two wheels at the front and a single track drive mounted under the rear. The operator sat on the plough that was hitched behind and drove the beast through controls that were mounted on long rods extending rearwards to him. The tractor itself weighed 2½ tons. Ferguson and Sands also worked with the Overtimes, which they knew well, Samsons, and International Harvester Moguls. The only British tractors they came across were Saundersons, a make which had been launched in 1905 and which, initially, had been unable to compete with the Ivel. But by the war years, Saunderson Universal tractors had been much improved and, though they suffered from the same defects of unwieldiness and excessive weight as the American tractors, they were 'quite a good job', in Sands's words. In fact, all the tractors of that epoch had deviated from the precedent of lightweight simplicity set by the Ivel. Farmers complained constantly of the packing of the ground by these heavy tractors, especially those whose farms lay on clayey soils.

During their tour in 1917, Ferguson, with his great persuasive powers, did the talking at demonstrations, though he would afterwards pull on a pair of working gloves and grab a spanner to help Sands with adjustments. They lived a spartan existence, and often stayed in unexpected places. One of the people whose tractors they looked at was a wealthy linen magnate called Frank Barbour. Sands knew him from a previous visit, just before the war, when he had gone to service his car. On that occasion, Sands and Barbour had travelled by train together to where the car was. At the station Sands was surprised to see Barbour come out of the station with a look of pleasure and satisfaction on his face.

'I've had a bit of luck!' he announced.

'Has a horse come up then?' asked Sands, knowing that Barbour ran a string of race horses.

'No,' said Barbour, 'I've managed to get you an excursion ticket.' And having handed Sands his third-class excursion ticket, he made his way to a first-class compartment.

Having this experience well in mind, and knowing Barbour's habits over the careful husbandry of money ('He lived, he grabbed, he died,' said Sands wryly)—Sands was quite relieved when on the occasion of his second visit, it was suggested that he should stay in a hotel while Ferguson was a house guest.

Their work done, and when they were leaving, Sands asked Ferguson how he had been treated by Frank Barbour.

'Oh,' said Ferguson, 'that man leads a very frugal life. Where can you get a good meal round here?'

And if Ferguson complained of lack of food, he must have been fed a starvation ration, for he himself ate very simply and frugally throughout his life, at one stage even fasting a day a week because he considered it beneficial to his health. People often remarked that he hardly seemed to eat enough to supply the needs of his spare, neat frame, let alone to support his electric personality.

The important aspects of the tour, however, were that Ferguson and Sands contributed a great deal to the success of the ploughing campaign, and at the same time, they gained a good insight into the tractor and plough design of the era. What Ferguson saw, particularly of plough design, appalled him. They were complicated in their construction—hundreds of parts all bolted together like a Meccano set—and they were extremely difficult to set to do good work. Indeed, the ploughs presented far more problems than did the crude tractors of the time, and shortly before his Irish tour, Ferguson wrote to the *Implement and Machinery Review*:

'I can assure you,' he wrote on the basis of his experience with the Overtime, 'that the tractor itself offers very small difficulties; the adjustment of the ploughs and getting them to do really good work was by far my greatest difficulty, and I have no hesitation in saying that the ploughs are a much more serious problem to the country at the present time when ploughing work is urgent than are the tractors. Before I started the work I was of the opinion that the plough would be the simplest possible proposition, and I did not anticipate the least trouble with it. In the end I had to turn my attention to the plough myself, and it was two months before I was able to get enough experience to be able to adjust any four-furrow plough to give reasonable service.'

What he saw during the Irish tour confirmed his low opinion of the plough design of the era. Quite apart from the intricacies of setting them, their design and their attachment to the tractor made them dangerous to use. This was because designers were still thinking in terms of implements as conceived for use with draft animals: the tractor was nothing more than a replacement for a team of horses, and the ploughs were merely enlarged to allow the full exploitation of the extra pulling power of the tractor. The all-important hitching of the plough to the tractor remained in principle the same as when a team of horses were used, that is to say there was either a chain or a bar from the plough attached, by means of a single pin, to the drawbar of the tractor. The plough, which had its own wheels, was simply towed along. The wheels, as is the case with horse ploughs, were used as the depth-regulating mechanism. They could be wound up or down relative to the plough shares, and as the wheels rolled over the ground, their rising and falling over surface undulations kept the shares riding up and down with them so that they were always at a constant depth below the surface.

The effects of the single-point hitch on the tractor were disastrous, especially when an obstruction such as a tree root or rock was hit by a plough share. Firstly, whereas a horse team would immediately have come to a halt when the obstruction was hit, the situation with a tractor was quite different. For a tractor in motion has much stored energy in its rotating fly-wheel, and the heavier the fly-wheel the greater the stored energy. Therefore, when one of the monstrous tractors of those times was brought to a sudden halt (because one of the plough shares had struck against a rock or root) the stored energy still attempted to drive the tractor forward. But forward it could no longer go, anchored as it was by the plough fouled against an obstruction, and the only way that the energy could be absorbed was by winding the tractor round its own back axle so that the front reared into the air. Had the rear wheels of the tractor been able to spin, there would have been an outlet for the stored energy; but, at the moment of impact, wheelspin was made even more difficult, by a strong downward pressure on the tractor drawbar, caused by the plough attempting to rotate around its depth wheel as a result of the force against the shares. This force, by adding weight to the tractor rear wheels, gripped them even more firmly to the ground so that they could not spin; at the same time, the downward force on the drawbar aggravated the tendency of the tractor to lift its front wheels off the ground (see Fig. 1).

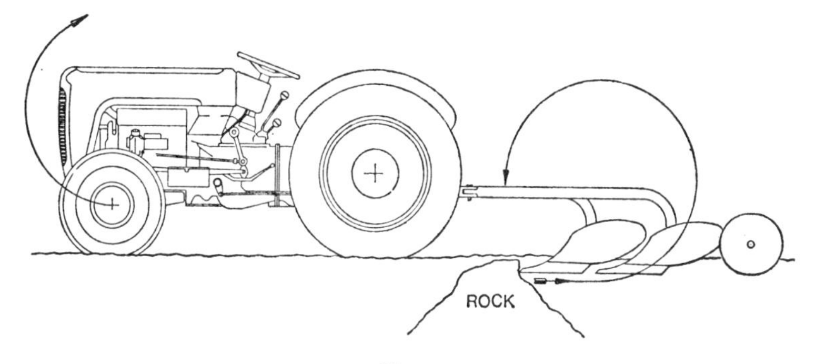

Fig. 1

The danger inherent in this tendency of the front of the tractor to rear up was obvious: it could tip over onto the driver and kill him unless he were quick to depress the clutch pedal and disconnect the drive to the rear wheels.

A secondary effect of the fly-wheel continuing to attempt to drive the tractor forward after an obstruction had been hit was that there was more likelihood of damaging the plough than if the tractive force ceased immediately upon impact. Partly to reduce plough damage, and partly to prevent tractors from rearing over onto their drivers, it became usual to insert a wooden pin somewhere in the drawbar so that when an obstruction was hit, the pin would shear and break the attachment between tractor and plough. But the wooden shear pins often snapped in normal conditions without encountering an obstruction. The driver had then to rehitch the plough to the tractor and insert a fresh pin. In very heavy conditions this might happen every 30 or less yards until finally, exasperated by the frequent stoppages and the need to clamber up and down from the tractor each time, the driver would drive a nail or other steel spike into the hitch in place of the wooden pin and get down to some uninterrupted work—but without the safety factor of the shearing hitch.

Apart from the fore-and-aft instability in the tractor, induced by work forces when ploughing, the ploughs themselves were often *laterally* unstable. This was dangerous because many ploughs had an operator sitting on them to make adjustments as they went along. Once at a demonstration when Ferguson was driving the tractor and Sands was on the plough seat, they were on sloping ground when the plough wheel rode over a tussock and the implement capsized, luckily flinging Sands far enough down the slope for it not to land on him.

In addition to being dangerous, the equipment was so cumbersome that a ploughman required considerable strength to make adjustments, and even to yank the rope that operated the lifting mechanism to raise the plough bodies out of work at the end of the furrow needed a strong man. This might seem unimportant, but it was certainly not to Ferguson. With his light physique, he had great trouble with some of the operations, and he found this irritating. He was also irritated by the excessive weight of farm tractors. The Overtime, as has already been pointed out, weighed over two tons, and this was light for the epoch. For example, the International Harvester Mogul weighed over 4½ tons; the Little Devil, made by the Hart-Parr Co. of Charles City, Iowa, weighed 3 tons; and the Omnitractor of London weighed 3¼ tons; yet none of these tractors developed more than 30 horse-power. Apart from the ill effects of packing the land, these heavyweights absorbed much of their limited power to propel themselves. Their weight also made them costly to produce. Most of them cost over £300 (or $1500), a large sum if one remembers the decrease in purchasing power of money that has taken place in the last half-century.

The heaviness of these tractors was made necessary by the hitching arrangement of the plough. This is explained if we remember that the forces generated by such a hitching arrangement during ploughing were, on a small scale, the same as those created when an obstruction was hit (see Fig. 1). In fact, as mentioned earlier, there was a constant tendency for the front of the tractor to lift, particularly when it was pulling hard. This was not only dangerous but also made the tractor difficult to steer. In order to stabilise the front of the tractor and give good steering qualities, designers added weight to it, often in the form of unnecessarily large engine-cooling systems. Alternatively, they extended the frame of the tractor forwards to render what weight there was more effective. Thus it was quite common to see tractors with their front wheels placed far out in front of the main body, giving the machine an ungainly praying mantis appearance. But this was perhaps a less harmful way of attempting to overcome longitudinal instability than was that of building extra weight into the front of the tractor; for extra weight increased the rolling resistance so that greater traction from the rear wheels was necessary to drive it. Since traction, or wheel grip, is largely a function of the amount of weight on the wheels, the designers were then forced to build additional weight into the rear of the tractor too in order to obtain the necessary grip.

The basic fault in the whole concept of tractor and plough design was that it in no way tried to exploit to advantage the various forces generated by the implement at work. In fact, it is true to say that those forces were a hindrance and created serious problems of instability. By adding weight, the designers were merely trying to suppress those forces, rather than attempting to use them as allies.

Ferguson had a gift for seeking out the fundamental issues when facing a problem, and he saw that radical rethinking was necessary if farm machinery were to be efficient, easy to use, and above all if it were ever to be so generally adopted that the grinding toil of farmers' lives could be alleviated. This latter point was one which was close to his heart, for he had suffered his share of that toil.

On his way back to Belfast with Willie Sands, their Irish tour of tractors complete, Ferguson was thoughtful.

'There must be a better way of doing the job,' he suddenly said to Sands. 'We'll design a plough.'

These words marked the beginning of what turned out to be almost twenty years of experiment, of hopes, of heart-breaking frustrations, and of unremitting thought and work. Finally, after showing a degree of perseverance possessed by few men, Ferguson was to launch his System of farm mechanisation. This System was based on novel principles which today are used in almost every farm tractor in the world.

6 *First Plough*

'It is no more possible to design a plough which would be suitable for use with various sizes of tractors than it is to design a cart which can be drawn by a donkey or a Clydesdale, or a body that would be suitable for all makes of car.'

The simple truth inherent in those words represented a novel idea when Ferguson laid them down as a guiding principle, for until then ploughs had been built for use behind any tractor of any shape or any size. And he culled another simple truth when he decided that a plough should become a unit with the tractor when it was hitched on: it should not be trailed along on a chain, like an afterthought. To be able to recognise such truths that others have missed, and exploit them, is one of the hallmarks of genius. And it is strange how, in retrospect, many of these truths seem so obvious that one is surprised they were not thought of earlier.

Since Ferguson was critical of the heavyweight tractors of the period and of their effect on the land, he decided to build his plough for the lightweight Ford Eros tractor. This was a tractor conversion of the Model T car and it was made by a firm in St Paul, Minnesota. The kit for carrying out the conversion consisted primarily of larger rear wheels mounted behind the normal axle, on an extension of the car frame, and driven by sprockets that took the place of the normal road wheels. There was also a larger radiator for cooling the engine at the lower speeds inherent in ploughing and cultivating. There was such a general shortage of tractors that these conversion sets for the Model T sold quite well; but the idea that the farmer could take his family to church in the car on Sunday and then convert it into a tractor for Monday morning was far-fetched, because fitting the kit would have taken him most of Sunday afternoon.

By December of 1917, Sands had translated Ferguson's plough ideas into metal: the new two-furrow implement was presented to the farming public at a demonstration at Coleraine.

'A new wheel-less plough, for which great claims are put forward, has made its appearance in Ireland,' said one report.

It was typical of Ferguson to 'make great claims', for as we have already seen in connection with his flying, false modesty was not among his characteristics. But he had every reason to be proud of his plough for the Eros, for it was a great advance over existing implements: it weighed only 220 lb., or about a third as much as other two-furrow ploughs, and Ferguson hoped to reduce its weight by a further 50 per cent when more suitable alloy steels were available. It was made with less than half the number of parts of other ploughs.

Most ingenious of all, however, was the way the plough was hitched under the belly of the tractor, and forward of the rear axle. Thus the effort of drawing it—the line of draft—tended to pull all four wheels of the tractor onto the ground, and banished the alarming tendency of the front of the tractor to rear up in heavy going, or when an obstruction was hit. The plough was mounted in conjunction with compensating balance springs which allowed the driver easily to lower and raise it in and out of work by means of a lever beside his seat. This lever was also used for setting the required depth of work, an operation that could be done while in motion. And since the mounting arrangement brought the plough very close to the tractor wheels, they carried the plough up and down with them over undulations in terrain; therefore the plough itself had no wheel for maintaining the implement at the preset depth. As final points in the plough's favour, the shares and mouldboards were designed to cut and turn the furrows gently and easily, thus giving it low draft. The compensating spring arrangement allowed the plough to ride over any obstruction encountered, but if a share should lodge firmly under a rock or root, there was a shear pin in the linkage that would snap and prevent more serious damage to the implement. The way the six spare shear pins were mounted in a special clip on the plough frame, so that the driver would not waste time rummaging around in the tool box when he needed one, was an unmistakable touch of its designer's obsession for order and efficiency.

At one of the early demonstrations of the plough in Ireland the shear pin was a source of embarrassment. The beams on the prototype implement were made of cast iron pending the arrival of the first production versions that were to be of cast steel; but Ferguson did not bother to explain this to his audience during his dissertation on the wonders of the plough and its safety features. This was a tactical error because a share struck a rock and the whole plough collapsed in a rubble of smashed cast iron; only the shear pin remained intact, to the huge amusement of the onlookers. The *post facto* explanation as to why the implement had thus disintegrated was less persuasive than it would have been had it followed a note of warning about the cast-iron beams during the introduction to the demonstration.

However, even if that particular audience went away amused and cynical, the plough was well received by the farming community and by the press. It was said to be a 'remarkable implement' that combined 'primitive simplicity with modern efficiency'. There were favourable comments on the fine ploughing it did and on the fact that it was easy to operate, by 'disabled soldiers, women or boys'. The first demonstrations of the Eros plough in England went well and the implement received further press accolades. The *Implement and Machinery Review* even said that they could confirm that it would do all its makers claimed it would. References were made to its 'striking characteristics' and it was even stated that 'the fundamental difficulties of attaching a plough to a tractor (had) been overcome on this machine'—a judgement that was totally erroneous as will be seen, but was indicative of the novelty that the plough represented.

That first plough introduced Ferguson's hallmarks of simplicity, efficiency and ease of use, some of the traits that were to make his invention famous. He later summed up his philosophy of engineering in the motto: 'Beauty in engineering is that which performs perfectly the function for which it was designed and has no superfluous parts.' Later this saying was printed on notices that were hung on the walls of the Ferguson engineering departments as a constant reminder to his designers. The marketing of the Eros plough was, however, a battle lost before it was begun, not because of any unsaleability of the plough, but rather of the Eros tractor. For during the period that Ferguson and Sands were designing and building the plough, events were taking a turn that was crucial to the whole question of tractors in British agriculture. These events concerned the new Ford tractor launched in North America in 1916.

Henry Ford, the son of a farmer and with lifelong roots in the land, had aspired to build tractors even before he built cars; and once he had the means he was constantly tinkering with experimental tractor models. Finally his engineers produced a design that was relatively satisfactory, and in 1916 the Fordson was launched. Its interesting feature was that it dispensed with a frame and had the engine, transmission and axle housings all bolted together like the tractors of today. However, most important was that it had been designed for mass production on an assembly line, a technique founded by Ford and developed to the point of bringing a model T car off the line every 96 seconds.

In April 1917, the British Board of Agriculture asked the Royal Agricultural Society of England (R.A.S.E.) to carry out tests with two Fordsons imported from Detroit. The government had been urgently looking for a tractor that could be introduced into British agriculture in large numbers. (They had even had a prototype built by the Ministry of Munitions with the intention of leasing government tractors to farmers, but it never went into production, probably because the Fordson arrived on the scene.)

The tests of the Fordson carried out in Cheshire by the R.A.S.E. on April 24th–25th, 1917, showed the tractor to be reasonably suitable for use on British farms, and Lloyd George's government wanted it produced in Britain at once. Ford generously made a gift to the British nation of the patent rights of the tractor and agreed to establish a manufacturing plant at Cork, Ireland. (He chose Cork because his family had originally emigrated to the United States from that town and he wanted to help it.) He must have undergone considerable inner conflict before agreeing to assist Britain with tractor production: he had an abhorrence of war, to such an extent that he organised the Peace Ship crusade in 1916, a Quixotic episode which brought him public ridicule. He invited prominent Americans to join him in a trip to Europe aboard an ocean liner he had chartered and they were 'to get the soldiers out of the trenches by Christmas'. These pacifists fought like bushy-tailed mongrels most of the way across the Atlantic—to the delight of the journalists on board who reported the events fully—and by the time the ship docked in Oslo, Ford was so exhausted that he quietly slipped away and returned to America. Despite the fiasco Ford remained a convinced pacifist and was angered by the Allies' refusal to discuss peace with the Germans. But his love of the land triumphed, even if increasing British food production was indirectly a war effort.

In December 1917, just as Ferguson's plough for the Eros was ready, it was announced that the Ford tractor plant at Cork was to be established and that 6000 Fordsons would be imported from America as a stopgap until the Irish-built tractor was coming off the line. This news, which was disturbing for Ferguson, also caused resentment among the few British tractor manufacturers: they considered that the British government had given Ford a singularly favourable position in the British market at their expense—a betrayal, in fact, of British industry. But in reality, even if the Fordson was far from being the ultimate in tractor design, it did have unique features that made it suitable for mass production. It was therefore the logical choice for the British government.

The Fordson was quite an advanced design for the time and, at well under 1½ tons, it was relatively lightweight. The first one to arrive in Northern Ireland, however, made a very bad impression at a ploughing demonstration and match, for none of the people operating it were competent. It was the laughing stock of the day and an Eros tractor fitted with a Ferguson plough easily won the contest. But such minor setbacks in no way altered the fact that the arrival of the Fordson cancelled any possibility of the Eros, or any other tractor, establishing itself in the British market. For the Fordson, manufactured with all Ford's flair for production-line techniques, was marketed in Britain at £250, well below the price of a Model T with Eros conversion, or for that matter the price of any other tractor.

The Fordson was still, however, a tractor conceived merely as a replacement for a team of horses, a source of tractive power for trailing implements behind it, and it therefore suffered from all the evil characteristics of earlier tractors. Indeed, it suffered from vicious longitudinal instability owing to its high drawbar, high-speed engine and heavy flywheel. When an obstruction was hit by the plough, the rearing action was so rapid that by the summer of 1918 a number of tractor drivers in Britain had been killed. So well known was this notorious trait of the Fordson that one company marketed a special device that would depress the clutch pedal as soon as the front of the tractor rose from the ground.

Ferguson did not greet the arrival of the Fordson, and with it the ruin of the market for the plough he had built for the Eros, with the ire one might have expected. Few opportunities escaped him and the advent in large numbers of the Fordson represented opportunities for the opening of an even larger market for ploughs. He greatly admired Henry Ford and, above all, he was acutely aware that his own dream of improving farm machinery and thereby making the farmer's life easier depended, in the long term, on having that machinery mass produced cheaply. The involvement of Ford's resources and know-how would be one means to that end. When, in the autumn of 1917, Ferguson first heard about the possibility of the Fordson being built in Britain and learnt that Charles Sorenson, Ford's right-hand man for forty years, was coming to London to discuss the matter, he and Sands snatched up their plough drawings and rushed to London to see him.

Ferguson's opening remark when they met in the Ford offices was typical of a tactic he frequently used.

'Your Fordson's all right as far as it goes,' he told Sorenson, 'but it doesn't really solve any of the fundamental problems.'

Sorenson did not receive the comment with enthusiasm, for it referred to the latest Ford brainchild. But at least Ferguson had gained his full attention and he quickly went on to explain that the only means of achieving efficient farm mechanisation lay in equipment designed on the unit principle—that is to say the implement becoming part of the tractor when it was hitched on, but being readily detachable again. Sorenson agreed that this would indeed be the ideal state of affairs but that it was difficult to achieve. Ferguson then unrolled his drawings of the Eros plough, whereupon Sorenson's hostile attitude changed to one of keen interest, and Ferguson's persuasive gifts won the day.

For his earnestness, his enthusiasm and that supreme confidence that he was right were almost impossible to resist—as they were for the whole of his life except in cases where his assertiveness aroused such initial anger and antipathy that they became insuperable barriers. But this was not the case during that first meeting with Sorenson. Sorenson encouraged him wholeheartedly to continue with his plough development work. He lived to regret the encouragement he gave.

'Had I foreseen the consequences of that meeting, I would have avoided it!' he wrote later.

But however much of an opportunity the introduction of the Fordson into Britain represented, it still left Ferguson with the problem of disposing of his stock of Eros ploughs. Willie Sands designed a clever modification to enable it to be used with the Fordson, but it did not perform to either his or Ferguson's entire satisfaction. Ferguson wanted to redesign the plough completely for the Fordson, and it was over this stock of Eros ploughs and his desire to design a new one that he first ran into difficulties with the co-directors of his company. There was a feeling among some of them that a motor business such as Harry Ferguson Ltd. had no right to be meddling in the field of plough research and construction. They were resentful of the money that it was costing and they tried to persuade him to give it up. Over the ensuing years this resentment waxed and waned repeatedly, and naturally it waxed to its heights whenever there had been some technical or commercial reversal that made it especially unlikely that the agricultural engineering research would lead anywhere. On one later occasion there were manœuvres to oust Ferguson altogether from the business. All this made his work even harder; the technical difficulties and frustrations alone would have been sufficient to daunt most men, without the hindrance of supposed supporters attempting to put a stop to the work. To their great credit, neither McGregor Greer nor Williams ever lost faith in him, for they were able to share his vision of the benefits that efficient farm mechanisation could bring.

With the stock of Eros ploughs sold, work began in earnest on another designed specifically for the Fordson. The result was a second beautifully engineered plough with a novel hitching arrangement. It was coupled to the tractor by means of two parallel sets of struts one above the other, with the lower roughly in the position of the conventional drawbar. The arrangement was named the Duplex hitch. This hitch overcame, at one stroke, many of the difficulties and dangers of ploughing with a plough trailed behind a Fordson tractor from a single hitch point. We have seen how a Fordson tended to rear up when the going was heavy or when the plough fouled an obstruction, and we know that part of this rearing tendency was caused by the pressure against the plough shares creating a rotationary movement on the plough and exerting a downward force on the drawbar (see Fig. 1). The Duplex hitch in effect put a strut (leading from a pyramid built into the plough to the back of the tractor) in the path of that force and led it through the tractor so that it exerted its pressure downwards on the front of the tractor, thus stabilising it. The tougher the going, the greater the pressure, and it was quite impossible for the tractor to rear over backwards onto its driver.

Secondary advantages of the Duplex hitch were that it enabled the weight of the plough to be reduced to a minimum; it was built of special alloy steels for this reason. Traditionally, in the days of trailed ploughs, it was mainly the weight of the implement that forced the shares to penetrate the soil. But the geometry of the Duplex hitch, with its struts free to pivot at both ends, was such that the lines of draft set up by the hitch pulled the plough down to its working level and tended to hold it there. As in the case of the Eros plough, the Fordson plough had no depth wheel to help it follow ground undulations: it was instead coupled as close as possible to the back of the tractor so that it would follow the tractor wheels. This was not too successful, but overall the plough and the Duplex hitch were a clever semi-rigid arrangement that integrated the plough with the tractor on the unit principle; yet it gave the implement sufficient independence, through the linkage pivot points, for the front of the tractor pitching over uneven ground not to see-saw the plough up and down as well.

Having patented this hitching arrangement, Ferguson decided he should show it to Henry Ford. He therefore wrote to Sorenson to ask whether he could come to Detroit to give a demonstration.

Sorenson replied cordially to say that they would be pleased to see the plough.

Ferguson planned to take Sands with him to Detroit; later in his married life, and when funds were no problem, Maureen usually accompanied him on business trips, but on this occasion she had recently given birth to a daughter, Betty, their only child.

7 *First Contact with Ford*

Ferguson considered that the coming demonstration of his plough to Henry Ford would lead to vast opportunities and he excitedly helped to load the implement onto the truck to take it to the docks—during which process, dashing around with more exuberance than purpose, he dropped a balk of timber on Sands's toe. Sands, dourly disapproving of such high excitement overtly expressed, hobbled away in a silent huff and had to be coaxed back to the enterprise. And the ship on which they travelled, the *Cambrian Head*, had barely set course for Quebec before Ferguson began calculating how many ploughs such a freighter could carry, and at what cost. He repeatedly dragged Sands around the ship measuring the holds, taking notes and querying the officers on ship-operating costs. He was visualising his ploughs coming out of some Ford plant in their hundreds of thousands and seeing a fair proportion of them jamming the holds of such freighters. In such moods of enthusiasm Ferguson was like an incandescent squib, shooting from one thing to another, fizzing and sparking with energy. He was irrepressible and irresistible, except perhaps to the dourest Ulstermen like Sands.

They landed in Quebec from where they were to take a train to Windsor, across the river from Detroit. The crate containing the plough was loaded into the luggage van and Ferguson and Sands installed themselves and their personal luggage in a compartment. When the train stopped at Toronto, Ferguson announced that he was going to check that the plough was all right. A moment later a guard came through the compartment and told Sands that he was sitting in a carriage that did not go to Windsor and that he had better change quickly. Sands snatched up all the suitcases and got speedily out onto the platform, but at once the train 'gave a whoop', in his words, and steamed away, leaving him standing forlornly with the luggage. His sense of bewilderment and slight panic at being so abandoned were still evident when he recounted this anecdote almost half a century later. During the four-hour wait for the next train, he walked backwards and forwards near the station, never daring to go too far.

When he finally reached Windsor at about 10 p.m., he found a perturbed Ferguson waiting on the platform. He rushed up to Sands, radiating relief.

'Oh am I glad to see you, Willie,' he said, 'I gave that guard hell.' To the phlegmatic Sands, for whom the worst part of the experience was now over, his joy seemed slightly overplayed until Ferguson said, 'All my money's in my suitcase you know.' And it was—in gold sovereigns in order to obtain a better exchange rate—and all of it borrowed from Harry Ferguson Ltd.

Dearborn, the Ford headquarters, was then nothing more than a seedy village with dirt roads, its only feature of interest or importance being Ford's Rouge River plant, the hub of the gigantic Ford empire. In that period, Ford was beginning to build up the Rouge plant into one of, if not the, most complete manufacturing complex in the world. He wanted to free himself of dependence on outside suppliers. Thus cargo ships brought iron ore from his own mines to the docks he built at the 'Rouge'; his own smelting and rolling mills produced the steel he needed for the forges and stamping plants that made the various castings, sheet metal, and other parts of his cars. He even had his own glass plant. And still today, the Rouge plant is one of the only places in the world where you can see the iron ore arrive at one end and cars emerge at the other.

Ferguson and Sands were well received by Sorenson at the 'Rouge' where their immediate task was to have steel beams made for their plough. (There were still no steel-casting facilities in Belfast and the plough they had brought had bronze beams.) Soreson put every facility, including a tractor and field, at Ferguson's disposal; and he and Sands prepared their demonstration with meticulous care.

A day was arranged for Ford and Sorenson to come and see the plough at work; they listened as Ferguson extolled its benefits and explained its hitching arrangements, and they watched while he ploughed a few rounds. Ford was visibly impressed, but he made the mistake of misjudging Ferguson's character.

'We could use a guy like that. Hire him, Charlie,' he said to

Sorenson. For Ford considered that Ferguson was merely an impecunious machinery salesman. Indeed, by Ford's standards, he was a very small business man, but he was sufficiently established and prosperous—and independently minded—not to be open to job offers.

Sorenson made him an offer then and there, but Ferguson courteously refused. Sorenson raised the salary figure three times, but Ferguson would only talk about the possibility of Ford manufacturing the plough. Ford himself was so piqued by the refusal of excellent job offers, that he told him to take his plough elsewhere. But before doing so he asked whether he could buy some of Ferguson's patent rights. This offer was also refused with a smile. Both men had met their equals in stubbornness and, on that occasion, they parted without having established anything other than mutual respect. And Ford left the door open by telling Ferguson to keep in touch and to continue his work.

The Ferguson plough was still not at the level of perfection to which its designer always aspired. The trouble was that it did not maintain its pre-set depth of work well, failing to follow the contours of the ground properly and therefore calling for vigilance and continual adjustments from the driver. Admittedly such adjustments could be made from the tractor seat and while in motion, unlike many ploughing tackles that still required a man on the plough, or a stop while the driver got down from his seat to make the adjustment. From the start, Ferguson had known that it was necessary to couple the plough as closely to the tractor as possible, and this he had done within the limits set by the design of the tractor. But even this position was not close enough to the back wheels for them to act effectively as depth wheels for the plough. After the demonstration to Ford and Sorenson, he decided to return to Ireland and to try to improve the design.

Back in Belfast, Ferguson ran into a second storm with some of the other directors. They were disappointed that his trip to Detroit had produced no concrete result and no prospect of revenue, and again they tried to make him give up all experimental work with ploughs. For several weeks the future of his efforts with ploughs hung in the balance, and Willie Sands, aware of the tensions and disputes, decided that the time had come to set up his own business. He therefore resigned from Harry Ferguson Ltd. and established, with his father, a small motor repair shop in Lisburn Road, Belfast.

Ferguson, as pugnacious an Irishman as ever lived, won his battle to continue with his plough experiments and accordingly set about trying to solve the problem of uneven depth of work. With Sands's departure, his chief helper was Archie Greer. Greer—no relation to McGregor Greer—was a pattern maker by trade and he had been working with Ferguson and Sands on the ploughs since 1917. He was a staunch member of the Plymouth Brethren sect, austere both in his habits and looks. He was convinced that God had sent him to work for Ferguson with the purpose of converting him from his appalling atheism. He had some extraordinary leisure interests. For example, he set about learning Russian from a refugee who had settled in Belfast, and after a few years is reputed to have spoken it quite fluently. He was an accomplished self-taught pianist and he also learnt shorthand and typing. He never married and he lived with his mother until her death. His sister and her husband lived next door, and yet another of his hobbies was making her clothes.

But to return to the plough, the obvious solution to the depth control problem was to fit a depth wheel on the implement so that it would faithfully follow the ground undulations, as had the trailed ploughs that Ferguson was trying so hard to improve upon. However, the fitting of a depth-wheel on a unit plough would immediately destroy one of the major advantages of the unit system. To understand this, we have to reconsider the whole question of weight and its role in increasing wheel-grip or traction: early tractors had much built-in weight on the front to stabilise the tractor when the draft forces of the plough were tending to make it rear up, and on the back to give the tractor extra wheel-grip or traction. The Ferguson unit ploughs, by virtue of their hitching arrangements, immediately overcame the tendency for the tractors to rear, but there was an equally important advantage following the elimination of the plough depth-wheel: without this wheel, the weight of the plough, in work or out, was always being carried on the back wheels of the tractor, thus giving them more grip or traction. Obviously, then, a tractor like the Fordson using a Ferguson unit plough was not only safer because it could not rear over backwards, but it also gained considerable traction because of the plough weight it was always carrying. Therefore, to put a wheel on the plough again and have it bear the plough's weight, would have destroyed a good part of the advantage inherent in the unit system—that of so-called weight transfer.

Ferguson and Greer spent much time and effort trying to overcome the problem of depth irregularity without fitting a depth wheel to the plough, and in 1921, they took the plough to the Fordson tractor plant at Cork to carry out further tests with some of the Ford staff. One of the two people detailed to help them was a young man called Patrick Hennessey. Hennessey, who had been shell-shocked and a prisoner-of-war in Germany and for whom working on the tractor production line was considered therapeutic, was skilled in tractor handling and went out to the field on many occasions to help with the plough. Hennessey, who was to rise to be chairman of the Ford Motor Company of Great Britain, receive a knighthood, and play a leading part in some of the dramatic episodes of Ferguson's later life, remembered well the field work carried out with Ferguson in 1921. He remembered in particular that there were some blazing rows. For Ferguson had fixed ideas about plough design, and if anyone tried to shift him from these, or was openly in opposition to them, the scene was set for a rousing argument.

In the event, no amount of tinkering with the geometry of the Duplex hitch, or the addition of an arrangement of springs, solved the problem of irregular depth of work. In the end, therefore, Ferguson was forced to fit a depth wheel on the side of the plough, much as it was against his principles to do so. He believed that even without the advantages of weight transfer from the plough to the tractor, the Fordson would still be able to cope. For, after all, no other plough was giving weight transfer either and the safety advantages and neatness of the Duplex hitch were in themselves important improvements.

8 *The Passing of the Depth Wheel*

With his plough as good as he could make it, or so he thought, Harry Ferguson decided that he must again go to the United States to find a manufacturer for it. Incredible as it may seem in retrospect, even by the early 1920s British farmers had still not decided that the tractor had come to stay. Indeed, the opinion was openly expressed that though tractors might have had some merit during those desperate war years, their continued use in peacetime was not warranted. Yet in 1920 in America, Ferguson's and Sands's eyes had snapped wide with amazement at the sight of five or six tractors all ploughing together in a single field. Obviously, therefore, America was where the real future lay for farm mechanisation.

On his second visit to the United States he was accompanied by his close friend John Williams who, although he had lost the use of an arm when he was shot down over France, was as active and gay as ever. Unfortunately, not much is known about that trip except that they travelled far and wide and lived in some discomfort. Funds were low and they could only afford to stay in the cheapest of boarding houses. Finally, in Bucyrus, Ohio, in May 1922 they showed the plough to John Shunk, the owner of quite a large blacksmith business, and he offered to put it into production. Ferguson and Williams made quite an impact in Bucyrus and when, at the end of May 1922, Ferguson and Shunk signed their agreement, the local Chamber of Commerce gave a banquet to mark the occasion.

At that banquet, Ferguson declared that a thousand ploughs would be built by the autumn, and that the first full year's production target should be 50,000 units. He added that every plough manufacturer would have to adopt his ideas and that perhaps half a million ploughs would be made at Shunk's plant. Indeed, he was confident that, very shortly, the whole of the Shunk plant would be needed for plough production.

'Bucyrus plant to make wonder plow' was the headline in

the Bucyrus *Evening Telegraph* following Ferguson's speech.

But when it came to it, for reasons that are not clear, the Ferguson/Shunk agreement was never implemented. Sands believed that Shunk lacked the manufacturing capacity, but perhaps the statements made by Ferguson at that banquet had startled him out of his wits and, thinking he had a tiger by the tail, he decided to let go as soon as he could.

Not knowing that their agreement with Shunk would come to nothing, Ferguson and Williams returned triumphant to Belfast in June 1922 and gave a press interview. Rumours had been circulating to the effect that they had sold the plough rights to some American company for a vast sum, but Ferguson disposed of these by telling the journalist that indeed offers 'exceeding even the most optimistic rumours' had been made to him but that he 'had decided to refuse them because the possibilities in the implement (were) too great to be disposed of for any sum that could be paid'. He commented that he could make no production plans for Ireland because of 'the state in which he found the country'.

By this 'state' he was referring to the civil war that had flared up in Ireland in 1921 and which involved his personal interests. For one of the very first acts of violence that had sparked off the general strife was an incendiary attack on the branch that Harry Ferguson Ltd. of Belfast had by then opened in Dublin. The attack cost the company a lot of money, but Ferguson was quite delighted by the affair. He recounted with glee that shortly afterwards, when driving through Drogheda, a motorist he knew who happened to be passing in the opposite direction shouted, 'Holy ——! There's the man that started the war.'

He had not been back in Ireland for long when he was informed that John Shunk was in fact unable to undertake the production as agreed. Once again, therefore, he set sail for America to begin the search for an alternative source of manufacture. This time, in Mansfield, Ohio, he managed to interest a company called Roderick Lean, famous for its disc harrows. The company happened to be making a set specially for the Fordson, and to manufacture a plough which was also specially adapted to that tractor could obviously be highly profitable. Accordingly, the company signed an agreement to manufacture the plough, with some of the components being made by the Vulcan Plow Company of Evansville, Indiana.

Again Ferguson returned to Britain confident that all was well, but soon disturbing reports began to reach him from America; they confirmed that his doubts about having a depth wheel on the plough had been well founded. In certain conditions, when the going was difficult, the Fordson was suffering from wheelspin. So yet another series of attempts at eliminating the depth wheel, and at the same time maintaining regularity of working depth, were necessary if the tractor and plough were to give optimum performance.

Faced again by the problem that had plagued his plough ever since its inception, Ferguson went to see Sands in his workshop in the Lisburn Road. Luckily Sands had by now come to the conclusion that running a small car repair business was a lot of work for small reward—'People never paid you. They just lived on your credit'—and when he was asked to come back and help with the plough problem he accepted. I felt when talking to Sands (as I did in sessions amounting to almost twelve hours), that in fact there was more than his disillusion with being the owner of a small business that led to his acceptance of Ferguson's offer: he obviously loved the development work that was such a challenge to his ingenuity.

By the time Sands was back at work with Ferguson, Lean's production procedures were established and a number of ploughs had already been sold. This complicated the task facing Sands, for it meant that any solution that he developed would have to be a modification to an existing design—one that could easily be incorporated in production and, more important still, be made up as a simple kit and sold to farmers cheaply to modify the ploughs they already owned. Sands succeeded brilliantly: he evolved a floating skid that was mounted on the rear of the plough and ran in the furrow bottom. (The patent for the device, which was applied for in December 1923, shows a small wheel rather than a skid, but a skid was cheaper and was therefore chosen for production.) This skid was ingeniously connected through a series of pivotal links to the Duplex hitch in such a way that it reacted to the movement of the tractor wheels as they rose and fell over ground irregularities. If, for example, the rear wheels of the tractor fell into a hollow (or the front rose on a bump), tending to pull the plough deeper into work, the movement of the linkage forced the floating skid downwards and thus the plough upwards so that it maintained its original working depth. If the tractor rear wheels rose on a bump, the converse happened, the skid lifting and thereby lowering the plough so that again it maintained its original working depth.

This floating skid completely solved the problem of depth irregularity, but at the same time it had the tremendous advantage of carrying very little of the plough weight, leaving most of it to be carried by the tractor where its usefulness in providing extra traction has already been explained. And the skid fulfilled all the other requirements of cheapness and ease of fitting to existing ploughs, as well as that of being easily introduced into production. Sands was justifiably proud of the device and the way it overcame so many knotty problems at one stroke.

The depth wheel was a thing of the past on Ferguson ploughs. In recognition of this achievement, there was a little ceremony (presided over by McGregor Greer) in a field at Andersonstown, about three miles out of Belfast, where the team did its field testing. A plough depth wheel was solemnly buried; in America, Sands related laconically, wheels removed from Ferguson ploughs when the skid was fitted found a ready use as spittoons.

Hardly had the technical problems been overcome, however, when fresh and grave commercial ones arose: Lean was beset by financial difficulties during 1924, and then went bankrupt. It would have been so easy at that point for Ferguson to have given up the whole plough business in disgust. But he did not: in 1925 he and Williams set out for America to make yet another manufacturing arrangement.

On this occasion, however, his search was easier, for during the course of his various other trips to America, he had made contact with the Sherman brothers, Eber and George. Eber Sherman had been Ford's export manager and was still connected with the export business. In addition, the Sherman brothers were the main Fordson tractor distributors for the state of New York, and when they compared the features of the Ferguson plough with those of others made for the Fordson tractor, they realised that here was a potential money-maker that fitted perfectly into their own business. Accordingly, they offered to join forces with Ferguson in order to manufacture the plough.

By now Ferguson had accumulated some capital of his own from his previous plough manufacturing ventures and so, in conjunction with the Sherman brothers, a company called Ferguson-Sherman Incorporated was set up, in December 1925, with manufacturing facilities in Evansville. The first cheque ever made out by the company was dated December 21st of that year and was in the sum of $1. The payee was Harry G. Ferguson and in the left-hand top corner of the cheque, where there was a space for stating the reason for the payment, were written the words: 'Down Payment on Success of the Unit Principle'.

Ferguson, his wife Maureen, and Williams spent almost a year in Evansville helping to get the business running. The plough, that was soon being made in volume, was well received by farmers and sold readily. The Ford Motor Company were obviously interested in it too, for one of their employees, Bill Checkly, worked with the sales team that demonstrated the plough widely throughout the United States. At last things seemed to be established on a firm basis.

9 Ulster T.T.

Once the plough business was running satisfactorily under the control of the Sherman brothers, Ferguson returned to Belfast, where, with the plough problems solved, his small research team of Sands and Greer were working in other directions. The aim had always been to apply the unit principle to all types of cultivating equipment, not merely ploughs, so that the weight of the implement would be carried on the tractor. To have the weight of the implement constantly on the tractor called for the abolition of an implement depth wheel; and in the case of the plough, this was achieved by Sands's floating skid. However, the floating skid could not be applied with the same success to cultivators. ridgers and other implements, for the skid needed to run on a firm base such as a furrow bottom to be effective. The team therefore had long discussions about ways of obtaining some sort of universal system that would automatically keep all types of soil-engaging implements, that were tractor-mounted on the unit principle, at a constant working depth, however much the tractor pitched up and down over irregularities in the ground.

Fortunately, once his plough was in full scale production in the United States, Ferguson found that he was able to carry on his farm machinery research with less resistance from the financial authorities of his Belfast company. In fact, quite apart from the plough revenue, the Belfast and Dublin motor businesses were doing well. Harry Ferguson was largely responsible in person for their success, for despite his preoccupations with the plough, he found time to keep a guiding hand on matters. He was helped by his younger brother, Victor, who had joined the company and who looked after it during his prolonged absences. Victor had an easier personality than his elder brother and employer. He was less intense, more sociable, and shared none of Harry's abstemiousness: to the contrary, and later his brother fired him, ostensibly because of his drinking habits. Equally likely, however, is that Harry was jealous of the popularity Victor enjoyed.

When Victor (who founded the still existent Victor Motors after the break with his brother) was killed in a road accident, most of Belfast turned out for the funeral in a spirit of real sadness; had the people turned out similarly for Harry it would have been more with a feeling of respect and admiration than for the loss of a friend.

It may seem strange that Harry Ferguson succeeded in having such a successful business in view of his long absences. The secret lay in his knack of selecting good people and inspiring in them unswerving loyalty and team spirit. His staff dealt with all the day-to-day business matters, while Ferguson devoted much of his energy to promotion of the motor trade in general. He saw racing and trials as one means of doing so and was instrumental in persuading the Ulster Parliament to pass a bill allowing the use of public roads for competition. He himself participated frequently in competitions, and with considerable success.

His own taste in cars was flamboyant and for some time he drove a daffodil-yellow and foliage-green Austin 16. Unfortunately, the same colour scheme was chosen by a customer whose main pastime was consuming large quantities of alcohol in convivial company, with the result that his car was usually to be seen parked outside some public house. Ferguson's friends and other customers, well knowing his abstemiousness, took to teasing him about the number of times his car was parked outside pubs.

'Will you either get that man to change his drinking habits or change his car,' Ferguson said to his staff in exasperation.

The business was still being run with all the obsessional attention to detail and order that had been its main characteristic during its early years. His salesmen had to be immaculate at all times, and woe betide them if they put their hands in their pockets. Hugh Reid, who originally joined the company as a journeyman draughtsman in 1918 and who today jointly owns the business with Joe Thompson (who started as an apprentice with Ferguson), recalls that on a number of occasions when he was deep in conversation with a customer he forgot himself and lounged comfortably with his hands in his pockets. Several times he heard a quick step behind him, his wrists were gripped and his hands lifted firmly out by his employer. It could never be said though that Ferguson expected of others what he himself was not prepared to do. His appearance was always immaculate and he was so neatly built that his diminutive height of 5 feet 6 inches was not particularly noticeable. He must have been conscious of it, however, for he wore shoes with a built-in half-inch rise. He was very interested in clothes and often gave an employee wearing a new suit a careful scrutiny. If the suit met with his approval he would say so, but he would also say so if it did not. He would also tell his staff to straighten a tie or fold a handkerchief differently, or comb their hair, always working on the principle that if someone worked for him he was representing him and must therefore behave and appear as he wanted him to. He carried this paternalistic approach right through his business life, even when he was the employer of several thousand people. His only concession to informality was the habit, probably acquired in the United States, of taking off his jacket in the office, but when he did so, he always exhibited a perfectly laundered shirt.

He could be humorous in relations with his staff. When Hugh Reid went to see him in his office to discuss some particular point he often found it easier to explain something by making a sketch on a piece of paper. That piece of paper was usually a letter snatched off Ferguson's desk, until one day, when Ferguson saw him appear in the doorway, he made an elaborate and theatrical gesture with his arms, gathering all the letters and papers towards himself and saying, 'If you're going to make any drawings, Hugh, bring your own paper!'

Following the stag hounds by car was his only pastime entirely divorced from work. When Patrick Hennessey once asked him what his hobbies were—during one of their conversations that did not impinge on farm mechanisation and when the atmosphere was therefore cordial—he replied,

'Taking the stag.'

'What do you mean?' Hennessey asked.

'Well, a stag that is being hunted sometimes ends up at bay in a farm yard or a stall, and I then go in with a rope and tie it up.'

This pastime must have called for considerable physical courage, but courage, physical and moral, was a quality he never lacked. Considering his relatively poor eyesight, some of his exploits in car speed trials in the 1920s also compel admiration.

Ferguson was passionately interested in motor racing, quite apart from commercial reasons he had for wanting to see it flourish in Ulster and of the benefit he considered it would bring to the 'imperial province'. The first important result of the bill that he helped through Parliament, was that a motor-cycle Grand Prix was held in Ulster in 1924, but he was determined to launch a major car race in Ireland. Since, as one journalist wrote, 'no lukewarmness or difficulties ever dampened his ardour', he persuaded Segrave and Lee Guinness, famous racing figures of the era, to come to Ulster to look at a number of possible courses, but none of them met with their full approval. He continued to push and pester in all directions and the newly formed Ulster Automobile Club agreed to run a car race over the motor-cycle Grand Prix course. The arrangements were completed and all seemed set when a competitor was killed in practice and the event was cancelled. The death of the unfortunate driver and the cancellation of the race brought great relief to Hugh Reid, however. For Ferguson was himself entered for the event and, having asked Sands to ride with him as mechanic and having received a blunt refusal, he had persuaded Reid to 'volunteer'. Thereafter Reid had been living in increasing trepidation as race day approached.

Ferguson continued to search for better courses and to importune anyone whom he thought could help. Finally, when many people had succumbed to his bubbling enthusiasm and were backing him in principle, the Royal Automobile Club agreed to organise a race if the Ulster government invited them to do so. Ferguson hustled back to Belfast to make sure that the invitation was in fact issued. It was, and thus the famous Ulster Tourist Trophy races were born.

The series of races, which began in 1928, brought the most famous cars and drivers of the world at that time to Ulster, in an era when it seems motor racing was more colourful than it is now. There was more hilarity and rather less death than today, even if the Ulster T.T. was finally abandoned on the Ards circuit after 1936 when a Riley went out of control on a fast left-hand bend in Newtonards, hit a lamp-post and ricocheted into the unprotected crowds, killing eight and injuring fifteen.

In the intervening years from 1928 to 1936, however, there were spectacular but harmless excursions into potato fields and rhubarb patches along the course, plus a few more serious accidents, but in general there was some thrilling racing. The T.T. became the most important occasion of the year in Ulster with at least half a million spectators turning out to watch, wet or fine. And Harry Ferguson loved the event that he had done more than anyone else to launch. His garage was the headquarters of at least one team of cars each year, but he rushed around all the other garages as well to see the cars and talk to the teams. He seemed to be in every garage at once, offering help and advice. He was particularly liberal with the latter, but since most of it was unrequested it generally went unheeded. He was so charming and engaging, however, that even if his advice and comments were uncalled for, nobody could take exception to him.

One of the regular and amusing features of the T.T. races was a row between Ferguson and the police or course officials. For he never had the right pit or paddock pass, doubtless expecting to be recognised and admitted on sight wherever he wanted to go. Inevitably someone did not recognise him and, having demanded entry and been refused, probably by an Ulsterman just as stubborn as himself, Ferguson became enraged. His voice went higher and higher and faster and faster as he got into his stride. One one occasion, some of the drivers expected him to be arrested, so heated was his row with the police.

The dances and banquets that followed the races—a banquet after the first race was actually given in Ferguson's honour—were the very rare occasions throughout his life when he overtly enjoyed himself. At one function there was an inflated figure of Monsieur Bibendum, the Michelin man, and Ferguson was seen puckishly prodding the figure with a table fork in an attempt to deflate it.

10 *The Master Patent*

We left the Ferguson research team pondering ways of obtaining a system of farm mechanisation that could be applied to all types of implement; it had to be a system which would allow the implements to be easily coupled to the tractor on the unit principle, and one which would automatically maintain the pre-set depth of work despite pitching of the tractor over irregular ground. Sands's floating skid achieved this for the plough, but Ferguson wanted to develop something more general in its application. And so immediately after the patenting of the floating skid in 1923, the team turned its attention to the broader problem.

There were numerous discussions, often around a blackboard with each of them making rough sketches. They built models and tinkered incessantly with full-scale implements. Each member of the team had full opportunity to express his own ideas and try them, but Ferguson always laid down the principles of what was required and kept the work moving in certain directions. Then, gradually, as they discussed and experimented, a new line of reasoning began to crystallise and gather momentum, and that reasoning went as follows.

An implement cutting through the soil at a certain depth—say a plough or cultivator at eight inches—requires a certain force, or draft to pull it. Obviously that draft load will increase if the implement runs deeper than eight inches, and decrease if it runs shallower. Why not use that fact to control the depth of work automatically? The draft forces are sensed through the points at which the implement is attached to the tractor and, if there were a mechanism for raising and lowering the implement, it should be possible in some way to have that mechanism actuated automatically by changes in the draft load. So, if our implement cutting at a depth of eight inches is forced deeper into work by, say, the tractor front wheels rising upwards on a hummock, the immediate increase in draft would give a lift signal to the raising and lowering mechanism and the implement would be lifted back to the point at which the draft was as before, i.e. a depth of eight inches. The converse would happen if the implement ran shallow as a result of the tractor pitching over uneven ground: as the draft decreased, a drop signal would be transmitted to the raising and lowering device and the implement would be allowed to sink back to the point at which the draft was as before, i.e. a depth of eight inches. In each case, once the draft that corresponded to the required depth had been re-attained, the raising and lowering mechanism would return to neutral.

The only inherent disadvantage in this principle was that changes in soil type or condition could also change an implement's draft, but Ferguson and his team reasoned that within a given field there were seldom major changes. And in any event, the system should be so arranged that the driver could easily make a manual adjustment to alter the implement's depth should there be, for example, an area of heavier soil which, when the implement entered it and the draft increased, would cause the depth of work to become less than the eight inches elsewhere. A simple adjustment by the driver, as he went along, would increase the draft setting in order to re-establish the desired eight inches working depth.

The whole principle described here is now known as draft control. Ferguson applied for a patent for it on February 12th, 1925, and it was granted in June 1926. The patent was also filed in America. Without doubt this patent, which was entitled 'Apparatus for Coupling Agricultural Implements to Tractors and Automatically Regulating the Depth of Work' can be considered the master patent of the Ferguson System; and draft control is now so commonplace in farm mechanisation that one can hardly imagine a time in which it had not been thought of. Once again it was a question of recognising a simple truth, one which appears obvious in retrospect.

Establishing the principle of draft control was, however, a very long way from making it a practical proposition. The draft control patent of 1925 shows a number of ways in which the principle could be applied. In each case the implement was tractor-mounted on the unit principle, in much the same way as the Duplex-hitch plough, and had upper and lower links. The draft force would put the lower attachment points in tension and the top in compression as the plough attempted to rotate about its axis (see Fig. 2). Most of the applications of the principle, as laid down in the patent, were based on using variations in the tension load on the lower attachment points to actuate various forms of raising and lowering mechanism. In one case the latter is shown as a system using two electric motors; in another there is a mechanical arrangement of cone clutches; and a third shows a hydraulic system. Yet another application of the draft control principle, as claimed in the patent, used the varying load on the tractor final drive, caused by changes in draft, to maintain the implement at the required depth.

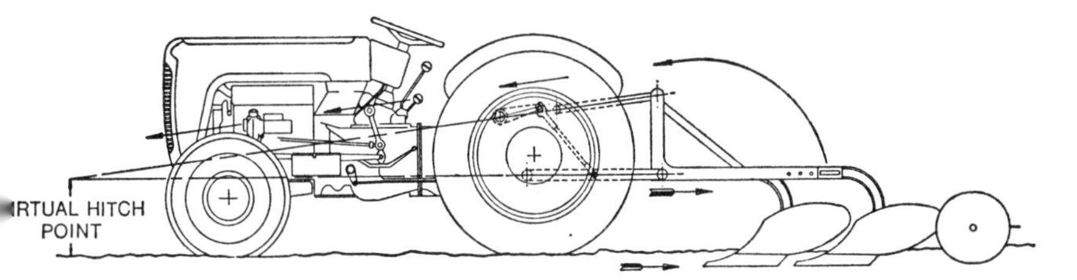

Fig. 2

Of all the applications covered by the draft control patent, the ones that seemed to offer most promise were the mechanical lifting arrangement (driven by a shaft off the tractor belt pulley and actuated through a pair of cone clutches) and the hydraulic lift. Both arrangements could be built onto a Fordson tractor, whereas the others required some fundamental changes in tractor design. Their adaptability to the Fordson was probably a main reason for Ferguson deciding to concentrate on these two methods.

The mechanical system appeared to be the easiest to develop, but the prototype they built encountered difficulties of wear on the cone clutches; and the ungainliness of the arrangement, with its special drive shaft from the tractor belt pulley, was certainly not the clean engineering that Ferguson always strove for. After very few tests, the mechanical system was put aside.

A hydraulic system for raising and lowering the plough

seemed to offer many advantages. In fact the hydraulic raising and lowering of a plough was not new: the principle had been used experimentally in the United States, but of course no one had tried to link automatic draft control to a hydraulic lift. As early as 1924 Ferguson had set Sands and Greer to work on a hydraulic lift that could be fitted to a Fordson tractor.

The actual lift mechanism was simple enough to design and make: it consisted of a cylinder with a piston, the connecting rod from the piston being attached to lift arms in such a way

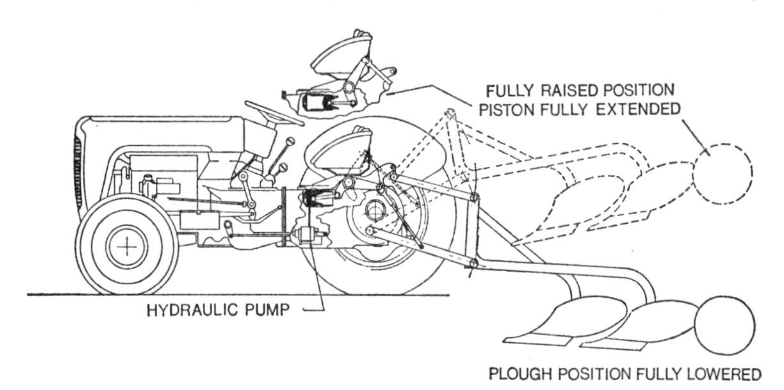

Fig. 3
Diagram to show principle of ram cylinder raising lift arms (as it was finalised in Ferguson System tractor)

that when oil was pumped into the cylinder the movement of the piston raised the implement. To lower the implement merely required the opening of a valve so that the weight of the implement could force the oil out from behind the piston and the lift arms would drop.

The main difficulty involved in the early stages was the design of a suitable pump and valve system, for to achieve automatic draft control required a sensing arrangement under which changes in draft of the implement at work would cause oil to be pumped into the cylinder, held there, or released according to whether the implement needed to be raised, held, or lowered to maintain its pre-set draft—and therefore its depth provided the soil remained constant.

The first line of attack was to have a pump that would deliver a variable amount of oil as the occasion demanded. The pump that they designed and made was a variable stroke swash-plate type with nine cylinders. Put simply, this pump had nine pistons arranged circularly in a pump body, the whole of which rotated. The ends of the small pistons protruded from their cylinders sufficiently to allow them to run against a stationary plate—the swash plate—the inclination of which could be altered, relative to the rotating pump body. As long as the swash plate remained vertical, no axial movement was imparted to the pistons as their ends travelled around on the plate; but if the swash plate was angled, the pistons would slide out of their cylinders and suck in oil as the rotation of the pump body increased the distance from the body to the plate; as the continued rotation reduced that distance again the pistons would be forced back into the cylinders, thus expelling the oil. The amount of movement of the pistons (i.e. their length of stroke and hence the amount of oil delivered) depended on the angle of the swash plate.

The pump was beautifully made with ball bearings to reduce friction to a minimum, but when it was put to the practical test it proved impossible to obtain sufficiently accurate control of the swash-plate angle through changes in the implement draft, mainly because moving the swash plate required considerable force. A more sensitive arrangement was needed, so they were forced to abandon the variable stroke pump in favour of a constant delivery pump with some means of circuiting the pumped oil back to the sump when it was not required. They at once designed and built single and double cylinder pumps, but both vibrated badly and were scrapped as unsatisfactory.

Month after month they experimented and tinkered with different types of hydraulic arrangement and pondered different ways of harnessing the changes in draft force to actuate the raising and lowering mechanism. They built a hydraulic lift onto the back of a Fordson tractor for their practical tests and took it to a farm at Andersonstown, just outside Belfast, where they spent whole days, in sunshine or in rain—more usually Ulster driving rain—adjusting, testing, adjusting again and testing again in a seemingly never-ending ritual. But, step by tiring step, progress was made.

Much credit must go to Sands, for he was nothing short of brilliant at providing solutions to the practical problems. He had a strange mannerism when confronted with a difficulty in the field: there were long silences while they stood looking at the plough or tractor, each deep in thought. Sands did not comment on anyone else's ideas for so long sometimes that it seemed he had gone dumb, or daft, or both. Then, quite suddenly, he would raise his hand as if he were about to salute and, spinning on his heel, make a complete revolution; at the end of it he would bring down his hand in a chopping, pointing motion. With a slow smile, he would then pronounce the solution. He was almost always right.

Most of the experimental work was carried out in a mood of serious dedication, but it was leavened from time to time. Once Ferguson asked Greer for a ruler to take some measurements, and Greer handed him a notched two-foot steel ruler of the type the firm issued to them. The notches in the ruler made them especially susceptible to breakage and Greer did not bother to mention that an inch had been broken off one end of his. Ferguson made a series of measurements, jotted down figures and tried to resolve the problem, but to his evident surprise nothing seemed to work out properly.

'By the way, Mr Ferguson,' Greer suddenly said, 'there's an inch broken off that rule.'

Ferguson, who never swore despite having given up the religious convictions of his youth, grunted in disgust and flung the offending ruler angrily over the hedge. Greer and Sands exchanged an amused look but wisely neither of them laughed outright. On another occasion Ferguson tried to bore a plough coulter-arm in the workshop, but the drill-bit embarrassingly emerged through the side of it. Again looks of ill-concealed amusement passed between all who had seen the performance, but again no one dared laugh outright. For they had respect for their employer; he was always 'Mr Ferguson' to them even after thirty years of working together.

In about 1927, commercial problems began to loom once again. Disturbing rumours filtered across the Atlantic from Dearborn:

Henry Ford was thinking of abandoning production of the Fordson tractor in North America, they said. Indeed, a vicious struggle had been going on in the tractor business in North America, and General Motors had been the first to withdraw because the rewards did not warrant the effort. Ford Motor Company was also finding the market difficult for their Fordson, and, in addition, they needed production capacity for the Model A car they were planning to launch. Evidently the Ferguson–Sherman plough business, which had been going well, was threatened.

Sure enough, in 1928, the blow fell: the Fordson tractor went out of production in North America, and with it the Ferguson–Sherman plough business was ruined. They had approximately 2000 ploughs in stock at Evansville and these represented a liability rather than an asset. In the event, they were not too difficult to sell because a fair number of Fordsons were imported into the United States from the Cork plant. Nevertheless, the Ferguson–Sherman plough business fundamentally collapsed with the discontinuation of the Dearborn-built Fordson.

The first consequence in the research programme was that Ferguson informed Sands and Greer that they would probably have to accept a cut in salary. Sands, who had foreseen something of the sort when the first rumours about the demise of the Fordson in North America had been heard, decided to leave; he purchased a Bean bus and set himself up in the field of public transport. Archie Greer stayed with the company, even though the future for the agricultural engineering research looked sour. For yet again, there was pressure on Ferguson to give it up.

The year 1928, though it brought him the realisation of his dream of establishing a major race in Ulster, also brought him a serious setback and many fresh troubles in the sphere of his research work. As usual, when faced by such adversity, and once he had overcome his first bitter disappointment, with the help of his wife he returned to the fray more determined than ever.

11 *Almost a Deal*

Despite the wrecking of his plough business, Ferguson still knew beyond doubt that the wide horizons opened up by the draft control principle he had patented in 1925 must be explored to the full; for an effective draft control system would be a real breakthrough in farm machinery design. But as usual after a major set-back, he needed to reconvince his fellow directors of the wisdom of continuing with agricultural engineering research; once again they grudgingly agreed to let him go on. One of the difficulties in connection with the research programme had always been that the accountancy section of the company used to debit any items for which they could not find another home to the research activities, thus making them a sort of bookkeeping rat-hole down which to throw all manner of odds and ends. Unfortunately, the odds and ends added up to quite a big sum and made it appear as though the research team was spending more than it was. In point of fact, Sands was always particularly proud that, considering the amount of original research work they were doing, their production of scrap was very low.

By 1928, hydraulic draft control was showing promise of becoming a practical proposition, but it was still a long way from perfection, and some inherent problems of implement control remained unsolved. To understand these problems, we must consider again the case of an implement being trailed behind a tractor from a single pin through a drawbar extending rearwards from the back axle. Quite apart from the aforementioned snags of such an arrangement (in particular the tendency for the tractor to rear up because of the forces pushing down on the drawbar and because of the line of draft running above the front axle (see Fig. 1)), there were others almost as inconvenient, even if less dangerous to the driver. For example, when working across a hillside, the rear of the tractor would naturally tend to slew downhill taking the implement with it and making it impossible to do a fully satisfactory job of tillage.

Equally important was the behaviour of the implement when the tractor was steered right or left. To take an example, if the front of the tractor was steered to the left, the first reaction was to swing the drawbar—since it was mounted behind the rear axle—to the right, carrying the implement with it. In other words, the implement did not immediately follow the steering of the tractor and this made it very difficult to use cultivators trailed behind tractors for hoeing between rows of plants such as sugar beet or potatoes. For every time the driver steered one way, the implement would initially swing the other and do an unrequired job of thinning among the plants—probably removing about two feet of each row at a time.

To overcome this problem, the simplest arrangement would have been to hitch the implement from the centre of the front axle so that it accurately followed the steering, or to mid-mount the cultivator under the belly of the tractor. Both these solutions, however, imposed limits of vertical movement on the implement.

Although the mid-mounting of cultivators on high clearance tractors with special wheels and front axles has since become commonplace in the parts of North America where maize or corn are predominant, Ferguson wanted to find a solution that would allow *any* implement to be coupled to the tractor on the unit principle with hydraulic draft control. For he was always aiming at a system of farm mechanisation that would be universally applicable—tractors and implements that would meet the needs of the farmer in the Kansas corn belt, the potato fields of Ireland, or the paddy fields of Asia. The implements would vary, but their method of attachment to the tractor and their control by draft forces would be standard. He therefore refused to be side-tracked by the possibility of having alternative hitching arrangements, and he determined to make his rear mounted system the basis for all tillage implements.

Returning to the Duplex hitch and Fig. 2, it will be remembered that the provision of two hitch points gave a completely different line of draft from that when a single hitch point was used. *Effectively* the draft was from a point where lines extended from the upper and lower links converged, in other words a virtual (or imaginary) hitch point other than the actual hitch points on the tractor. Ferguson realised that the same principle of an imaginary hitch point could be used to solve the problem of lateral control of the implement, thus eliminating the tendency for the implement to swing initially in the opposite direction to the way the tractor was steered. At the same time the necessary freedom of lateral movement could be retained provided the links had swivelling ball-joints at their connection points.

In 1928, he applied for a patent showing the practical application of the principle. It was nothing more than arranging the two links, or upper arms, that connected the implement to the tractor in such a way that their ends on the tractor were closer together than their ends on the implement. The result was to give a converging linkage (see Fig. 4) whose lines of draft met at a

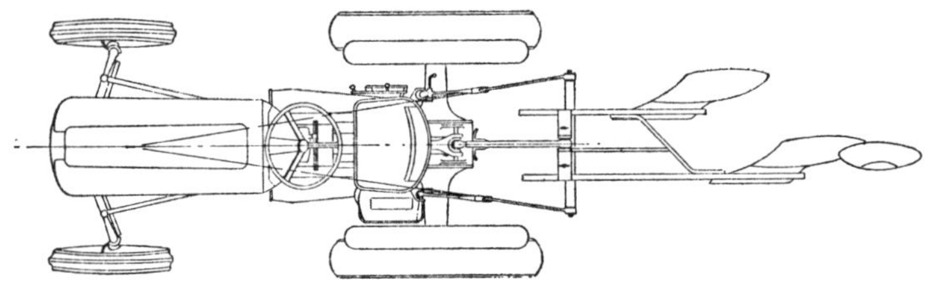

Fig. 4

point other than the actual connection point between tractor and implement.

This patent, in fact, was a natural extension of that covering the Duplex hitch, for it was an application on the horizontal plane of the type of geometry built into the Duplex hitch in the vertical plane. In the words of the converging linkage patent specifications of 1928, 'the line of draft . . . acts as from a point other than the actual connection between the two (tractor and implement) and when the tractor is turned the implement will swing laterally relative to the tractor about a point other than the actual connection between it and the tractor. Thus the advantages of the (Duplex) hitch described in my prior patent specification . . . are combined with accurate steering control over the implement.'

Even with the problem of the linkage geometry satisfactorily settled there remained others with the hydraulic system and the connection of the linkage to it so that draft forces would actuate it to raise or lower the implement as required. One of the most serious of these problems was heating and aeration of the oil. The four-cylinder piston pump that they were by now using pumped oil continuously, and when oil was not needed to raise the implement it was bled off through a 'leaky valve' arrangement. This continuous pumping through the 'leaky valve' heated the oil and aerated it; and air is an enemy of hydraulic systems, for its compressibility makes them spongy and sluggish.

Despite these remaining difficulties, the makings of a satisfactory arrangement were evident. The hydraulic system was built onto a Fordson with two upper links that raised and lowered the implement, and a single lower link in tension through which changes in the draft moved the hydraulic control valve to the raise, lower, or neutral position. This modified Fordson was demonstrated to influential people in Ireland and England in the hope of gaining support from a manufacturer who might put the system into production.

In the autumn of 1928, Ferguson and his wife visited America to see what could be done about the Ferguson–Sherman plough business and to make contact with companies that might be interested in the hydraulic draft control system. Allis-Chalmers showed some purposeful curiosity.

Not long after Ferguson's return to Belfast he received an unexpected visit from Willie Sands. Sands's bus business had prospered and he had just sold out to a larger company. With the nest-egg resulting from this sale, swollen no doubt by savings made possible by canny Ulster frugality, Sands had come to make Ferguson an offer.

'Mr Ferguson,' he said, 'I'll give you two years of my services for no salary. If in that time we get the system right and make a really good job, you can pay me back. And if we don't, we'll just say no more about it.'

The offer touched Ferguson deeply; he was an emotional Celt in many of his reactions. But he could not accept, and neither did he need to, because the co-directors of the company were beginning, at last, to think that perhaps this development work might be leading somewhere profitable. He therefore turned down Sands's offer but at the same time said that he would be truly delighted to have him back in the research team as a paid employee. Thus the Ferguson/Sands/Greer triumvirate was reconstituted for the last long offensive against the problems still existing along the path towards a fully efficient draft control system.

Concurrently with trying to solve these remaining problems, Ferguson continued to try to arouse interest in his system by demonstrating it widely in the hope that a manufacturer might incorporate it into their tractor. The first company to rise to the bait was Allis-Chalmers which, not long after his 1928 visit to America, took out a ninety-day option on the inventions. The option was never exercised, however, probably because the draft control system was really in need of more refinement.

The Rover Car Company, who were planning a tractor with rubber-jointed tracks, showed interest, and Ferguson sent Sands scurrying to England to carry out some tests with their prototype. To his disappointment, he found that the tractor was totally unsuitable for the incorporation of a draft control system; it was so short that it pitched violently over uneven terrain, and even a highly refined draft control system could not have kept the plough at an even depth of work as the tractor made its rocking-horse progress across the field. Said Sands, 'You had to hold on like grim death as you crossed furrows.'

The Ransome Rapier Company was also planning a tractor about that time and was prepared to incorporate the Ferguson inventions, but they were undercapitalised and the tractor was never produced. Another tractor, the Rushton, actually had special faces incorporated into its design so that the Ferguson draft control system could be bolted on.

There was, then, a fair degree of interest in the Ferguson inventions, almost all of it generated by demonstrations of the Fordson onto which the hydraulic system had been built. These demonstrations were part of a promotional campaign that was in some ways premature, for the system was still not perfect. And Ferguson knew so, even if his speeches at demonstrations gave the impression that here was the last word in farm mechanisation. There remained the problem of overheating and aerating the oil; and under certain conditions the system worked with a jerkiness that made the plough bob up and down slightly. This left an uneven furrow bottom, and super-critical farmers often failed to see beyond this. They did not appreciate that here was a system that allowed the implement to be raised and lowered effortlessly at the touch of a lever near the driver's seat; they overlooked the important fact that once in work (unless the soil type changed) the plough was kept at its pre-set depth automatically even while the tractor was pitching over uneven ground; nor did they understand the significance of the weight transfer arising from the abolition of the implement depth wheel. The slight jerkiness of the system and the bobbing of the plough blinded them to all else. But for those who really used their eyes and minds together, what they saw was almost miraculous in comparison with the crudities of most farm machinery of the time. To them, there lay in that admittedly imperfected system a very important potential. These people with the ability to grasp the implications were those whom Ferguson was attempting to reach. And if any of them lacked faith in his and his team's ability to perfect the system sooner or later, they were underestimating the man's powers of perseverance to see it through to the end. For he had a philosophy in this respect which was summed up for him by a saying of Sir Francis Drake's that he often quoted: 'There must be a beginning of any good matter, but the continuing to the end, until it be thoroughly finished, yields the true glory.'

Out of a sense of deep patriotism Ferguson would have liked to see his inventions built into a tractor manufactured in Ireland. The country's industry, and particularly its shipyards, were bogging down in the depression of the early thirties. A tractor industry could have been a fillip to the economy, but none of the companies with sufficient resources, such as Harland and Wolff, could be persuaded to take an interest. Another of the shipbuilders he approached was on the verge of bankruptcy and John Williams went to look at their premises. He came back to announce that the only tangible asset the premises had was the pigeons, to a value of about nine-pence each, roosting around the derelict buildings.

After a fruitless search in Ireland, they took their Fordson with Ferguson hydraulics to England. They went first to the Norfolk estate of Tim Birkin, the racing driver with whom Ferguson had become friendly as a result of the T.T. races. In Norfolk they carried out some more tests and demonstrated to one or two influential people. But the main purpose of taking the tractor to England was to show it to the Prince of Wales. After a certain amount of pushing and persuading by Ferguson, the Prince of Wales and the Duke of Kent agreed to attend a demonstration at Ascot, on a farm near the racecourse. They were openly impressed by what they saw and heard, even if the technical aspects of farm machinery were of little interest to them. Perhaps they dropped a word in someone's ear, for a few days after their visit, the Morris Motor Company sent a representative to Ascot to look at the draft control system and to assess the commercial possibilities of building a tractor incorporating it. His report back to Morris resulted in an offer to co-operate in the design and manufacture of a tractor with implements to match.

It seemed as though a commercial breakthrough had been made and Ferguson, delighted, immediately enlisted the help of the best lawyers to draft an agreement. This agreement was approved by all concerned, but about a month later, when everything seemed settled and when Ferguson was on the verge of leaving Belfast to go to England to sign it, Morris sent him a telegram cancelling it. Why this happened is not clear, though it seems probable that Morris Motors had come to the same conclusion as many other companies: the general economic situation at the time did not allow for expansion into new fields, and especially not into those of agriculture. For farming was generally looked upon by industrialists as primitive, hazardous and unrewarding, and farmers were considered second-rate citizens. Ferguson constantly and rightly proclaimed that agriculture was the most basic and important of all industries, and that it was inadmissible that it should remain primitive, but his words had no effect on the industrialists whose help and facilities he needed. The collapse of the Morris agreement, when it was so near to completion, disappointed him bitterly. 'But he didn't show it much,' said Sands. 'I suppose that was part of his policy too. What with all the frustrations he'd had, he was very used to them.'

12 *The Black Tractor*

It was obvious by the end of 1931 that no company was going to embark on a new venture of tractor production despite what Ferguson or anybody else told them. He had already tried to find an existing tractor suitable for the incorporation of his ideas, but there really was no such tractor on the market at the time. For the tractor had to have certain characteristics if it were to make the most of hydraulic draft control. It will be remembered that one of the prime advantages of eliminating the plough depth wheel was that of enabling the implement weight to be carried entirely on the tractor at all times, thus giving it extra wheel grip. To exploit this principle to full advantage required a tractor that was in itself light, a tractor that was not continually carrying unnecessary built-in weight, packing the ground and absorbing excessive power to move itself. Nothing really suitable existed: the Fordson built in Cork and by now being marketed from Trafford Park, Manchester, was the lightest tractor, at under 1½ tons, in volume production. The Austin had weighed about the same but had gone out of production in England not long before; a few were still being made in France. Other makes weighed far more: the Rushton, with an 18 hp engine, weighed about 2½ tons; one International Harvester with 9 hp weighed about 2 tons; Massey-Harris had 20 and 30 hp models weighing just under 2 tons and about 2¼ tons respectively; and the Hungarian Hofherr Schrantz single cylinder diesel of 30 hp weighed over 3 tons, monstrously heavy for its power.

It became clear to Ferguson that the only way forward lay in building a prototype tractor incorporating his own inventions, an ultra-lightweight machine which could ultimately be built very cheaply and which would be useful on the smallest farms as well as on the largest. With such a prototype in existence he might be able to find the support he needed to go into production. In 1932, he, Sands, and Greer began sketching out their ideas for the design of a tractor incorporating the hydraulic draft control system. It was to embody the best and most advanced engineering techniques, many of which Ferguson knew from his dealings with racing cars.

Within a few months, the general layout of the tractor had been decided and a draughtsman was needed to make the detailed drawings. The person selected was a Downpatrick farmer's son, John Chambers, a young man whose precision Ferguson much appreciated and with whom he became closely associated in later years. Chambers and Greer produced the detailed drawings for the tractor, and specialised companies were approached to make certain parts of the prototype. The obvious supplier of the gears, transmission and steering was the David Brown Company of Huddersfield, a family business that had grown from small beginnings in 1860 to be the largest producers of gears in Britain. Today, of course, the David Brown Companies are world famous, not only for gears but for their Aston Martin cars, Vosper patrol boats, and their own David Brown tractors. The latter grew from a single order for gears, transmission, and steering placed by Ferguson in 1933.

The engine chosen for the prototype tractor was a four-cylinder American Hercules giving about 18 hp. The hydraulic draft control system was an adapted version of the one that had been used as an external unit on the Fordson: it had been modified in that the three-point linkage for mounting the implement on the tractor had been turned upside down, that is to say that there were now to be two lower links and one top as opposed to vice versa on the Fordson. The lower links, which were to do the raising and lowering of the implement, were also to be the means of transmitting signals to the hydraulic control valve when there were changes in draft. The pump was to be built into the tractor, rather than being bolted on externally as it had been on the Fordson, but it was still to pump oil continuously, the excess being bled off through the 'leaky valve' arrangement.

As the tractor began to take shape, small and neat compared to others, the excitement in the team grew with it. Ferguson decreed that it was to be painted black, perhaps influenced by Ford's statement that black was the only suitable colour for cars, but more likely because of Ferguson's own liking for functional simplicity. He probably felt that immaculate black would be the finishing touch in setting off the clean and well thought-out engineering of his prototype. The 'Black Tractor' became the name by which this prototype was, and is, always referred to. It is now in the Science Museum, Kensington, a fitting place for it to rest, introducing as it did a completely new concept in farm machinery design.

The Black Tractor was complete by 1933 and it was immediately put on test with the various implements that had been designed for operation with a hydraulic draft control system. There was a plough, a ridger for potato cultivation, a row crop cultivator, and a normal cultivator or tiller. This tiller was an implement that had been developed sometime earlier by the Ferguson team; it had a patented arrangement by which the tines were spring-loaded, allowing them to kick back if an obstruction was fouled and then return, undamaged, to their normal position.

But the first field tests of the hydraulic system of the Black Tractor were disappointing: for the problems that had plagued the system as mounted on the Fordson lingered on. The hydraulic oil was still being heated and aerated despite some detail changes in the 'leaky valve', but just as serious was the bobbing which resulted in an uneven furrow bottom when ploughing under certain conditions. It seemed that having the draft sensing transmitted to the hydraulic control valve through the same links (the two lower ones) that also raised and dropped the implement might be part of the cause. Sands therefore thought that instead of using the *tension* loads in the lower links to give draft control, they might try to use the *compression* load in the top link to actuate the hydraulic raising and lowering mechanism. He duly cobbled up a conversion and tried it in the field. It seemed to give better results, but Ferguson was impatient to have the tractor in production.

'It looks promising, Willie,' he told Sands, 'but we can't involve ourselves in major changes now. We'll have to see if we can't get round the problem some other way.' Sands therefore hung his modification up in an out-house.

But no other way seemed to be leading in the right direction, and about six weeks later, in desperation over the bobbing problem and general jerkiness of the system, Ferguson asked Sands 'to bring that thing out again'. He made a series of tests to determine the compression loads in the top link, and it became evident that indeed the idea of using these loads—instead of tension forces in the lower links—to actuate the control valve of the hydraulic system was practicable. They decided to work on this basis in future: they modified their prototype, connecting the tractor end of the top link to the hydraulic system so that the control valve would be moved by the changes in compression load in the top link brought about by changes in draft.

Once again, the search for a manufacturer was on. It began with a series of demonstrations in Ulster that were as unproductive as had been those with the modified Fordson. Ferguson continued to maintain, to anyone who would listen, that manufacture of his tractor in Ulster could be of great assistance to the ailing economy of Ireland, but no one could be found to take up the idea. In the end he was again forced, reluctantly, to try to find a manufacturer in England.

The first demonstrations were given near the David Brown Park Works in Huddersfield, on land that is now the David Brown Companies' private airstrip. Among the people who attended were representatives of the Craven Wagon and Carriage Works of Sheffield, a subsidiary of the Thomas Firth and John Brown steel company. They were impressed and began to consider very seriously taking up the manufacture of the tractor. While they were mulling over the idea, the tractor was moved to Ascot to give demonstrations to influential people from London.

The Ferguson team by then included a thickset young man called Trevor Knox who had joined the company in Belfast as an apprentice on the automotive side. However, he joined when the Black Tractor was being planned and he was seconded to the tractor activities. Since he had little experience of farm work, despite being a farmer's son, one of the first things that Ferguson said to him was: 'Look, you're to take Saturdays off; you're going down to your father's farm and you're going to get behind a pair of horses.' For Ferguson considered that no one could grasp the import of what was being done without first going through the hard grind of unmechanised farm work and of ploughing in the traditional way. Knox later became a tractor demonstrator of skill second only to that of his employer.

One person who was interested in the future of the Black Tractor was David Brown (now Sir David). He was then barely thirty years old and had recently taken over as managing director of the David Brown engineering company founded by his grandfather. He had known Ferguson since the day about two years earlier when the Ulsterman had visited Huddersfield to place his order for the transmission, gears and steering of the Black Tractor. David Brown went to the Ascot demonstration:

'It was a good demonstration,' he recalls. 'All Ferguson's demonstrations were very good. He was absolutely first class; he was without equal at doing that sort of thing; of this there is no doubt. He was a superb salesman.'

Shortly after the Ascot demonstrations, the Craven Wagon Works of Sheffield decided to sign an agreement to manufacture the tractor; Ferguson would market it. Ferguson and his team moved to Sheffield in 1935 to oversee the preparations for production, but difficulties arose from the very start. Not unnaturally, the development team that had spent so many years developing the hydraulic draft control system, and the method of coupling implements to the tractor on the unit principle, was sensitive about even the slightest modification that was suggested; and Ferguson wanted complete control over the engineering and material specifications of the tractor. He was particularly adamant over the latter: the materials were to be the very best available, and he offended the Craven Wagon Works by requesting that American steel specifications should be used throughout. He thought the tractor would be more easily marketable in America if it was manufactured to American steel specifications, but as Sir David Brown commented: 'Firth Browns were in the steel business and claimed to know something about steel, so this was I think really the start of the trouble.'

During this 'trouble', Harry Ferguson took to dropping in regularly on David Brown in Huddersfield; and frequently he worked the conversation round to the point where he was quietly suggesting that Brown should take up the manufacture. Brown was in fact interested in doing so, for as the young and forceful new managing director of the family company he was expansion minded. But there were difficulties because Brown's father, Frank, was against the idea; he did not like tractors, or Ferguson, and bluntly said so. The other difficulty facing young Brown was that he did not know how matters stood between Ferguson and the Craven Wagon Company. But that was fairly easy to establish; he visited Charles F. Spencer, a director of Thomas Firth and John Brown, with whom he had been friendly for some years. In Spencer's office in Lower Regent Street, Brown asked him how they were getting on with preparations to manufacture the Ferguson tractor. Spencer made a gesture of despair.

'We can't get on with that fellow,' he said. 'Frankly, it isn't working out, and it isn't going to work out.'

He went on to elaborate the difficulties that were arising over Ferguson's insistence on having complete control over every technical detail. Spencer also objected to the lengthy letters crammed with advice that Ferguson used to write. (Ferguson used to tell the much younger Brown—in Sir David's words—that such letters were 'pearls of wisdom that should be stored and read and re-read'. But Charles F. Spencer didn't think they were such pearls.)

'Well,' said Brown to Spencer, 'what would you say if I were to take it off your hands. You have an agreement, would you be prepared to hand it over to me?'

'I would be delighted,' said Spencer, 'but I think you'd be a bloody fool.'

David Brown went away to think about it, and decided that he did indeed want to manufacture farm machinery; and so, despite opposition from his father and the gloomy warning from Spencer, Brown offered to build the Ferguson tractor.

Meanwhile, Ferguson was still disturbed about a serious shortcoming of the Black Tractor: using the single top link had considerably improved the draft control but there remained problems caused by the continuous pumping of oil under pressure and the tendency to heat and aerate it. On several occasions the problem was so severe with the Black Tractor that after half an hour of work the hydraulic system airlocked. The problem that had been dogging them for years really had to be beaten once and for all before the tractor was marketed.

By 1935, Harry Ferguson, who was just over fifty, was well on the way to being a chronic insomniac. Some of his best thinking was done during these sleepless periods, and one night the solution came to him with blinding suddenness. Again it was the simplest of ideas and reviewing it now it seems obvious: why not put the hydraulic control valve on the suction side of the pump, that is to say cut off the supply of oil to the pump when no oil under pressure was required to raise the implement? It would mean submerging the pump and control valve in oil but he could foresee no real problems. The control valve, which would be actuated by the draft forces in the top link, would be a sleeve running in a housing which was part of the pump. In that housing there would be two (inlet and outlet) ports or openings. When the sleeve valve was centrally placed it would block off both ports, thus retaining the oil in the ram cylinder and holding the lower links at their actual height. If there was an increase in draft as the implement went deeper, the extra compression load in the top link would slide the valve one way to uncover the inlet port and allow the pump to suck in oil and raise the implement back to the pre-set depth of work. If there was a decrease in draft the valve would be moved the opposite way to uncover the outlet port so that oil escaped from the ram cylinder back into the sump, thus lowering the implement. The automatic mechanism actuated through draft forces in the top link could, of course, be overridden by the lever next to the driver's seat when he wanted to raise or lower the implement on the headland or make an adjustment in working depth.

The next morning he told Sands of his idea. Sands was dubious at first, for he rightly said that after the valve had been moved to the inlet or lift position, the pump would complete half a revolution before any oil was actually delivered; and when the need to raise the implement had been satisfied and the valve returned to the central or cut-off position, the pump would still contain oil which it would deliver on the next stroke. He thought that this oil delivery slightly out of phase with the control valve movements would cause unevenness of depth. But he was wrong, for suction side control turned out to be the basic answer to the problem. However, some mild tendency towards bobbing of the implements under certain conditions was to remain for years, even when Ferguson System tractors were being sold in thousands, and it was only eliminated ultimately by detail changes to the control valve design.

Suction side cut-off was the last breakthrough needed to make hydraulic draft control a completely sound and practical proposition. A hydraulics engineer would say today that to cavitate a pump by cutting off its access to oil is theoretically not an ideal solution, but provided it worked and there were no ill effects on the pump, Ferguson would not have been deterred by such a theoretical point of view. He never had much regard for academic matters anyway, firmly believing that academic education, particularly at university level, 'polished pebbles and dulled diamonds'.

After the inspiration on suction side cut-off, the Ferguson hydraulic draft control system was at last reaching perfection. Only one additional feature remained to be added to it, and this came about almost by chance. One of the control valve arrangements built after adopting suction side control was laid out so that the valve had a great deal of axial movement, so much that a massive increase in compression load down the top link could push it past the position in which the inlet ports were uncovered until the *outlet* ports were also completely uncovered. The advantage of this accidental arrangement was immediately realised: if the implement hit an obstruction the violent surge in the draft load, in pushing the valve right past the outlet ports and uncovering them entirely, would allow the oil under pressure in the ram cylinder to escape very rapidly. This jettisoning of the oil from the ram cylinder back into the sump in effect dropped the weight of the implement, which until the moment of impact had been carried on the tractor through the oil under pressure in the ram cylinder. This release of oil pressure, and hence implement weight, freed the tractor wheels to spin harmlessly until the driver could declutch. The effects of the impact against the obstruction were therefore minimised, as was also the risk of damage to the implement or tractor. Ferguson called this arrangement 'overload release' and patented it. It was so effective that at demonstrations he would have a large steel bar driven into the ground in the path of the plough share. The tractor would then be made to approach at normal ploughing speed; spectators who had not seen the show previously winced at the impact and expected the tractor to rear up in traditional fashion; instead, however, the shock load down the top link firmly pressed the front wheels of the tractor to the ground, while the rear wheels, relieved of the implement weight by overload release, merely spun. Spectators were even more surprised when the driver reversed the tractor, raised the undamaged implement, drove forward a few feet and set it back in work as if nothing had happened. It was an impressive display, but it must not be overlooked that in addition to overload release, all the implements were made of high quality steel alloys which helped them to survive such treatment intact.

With the incorporation of overload release, and using suction side cut-off, the Ferguson System had reached a satisfactory level of technical and functional quality, even if refinements were added to it over the years. As Ferguson himself once described it in a letter, the System was like the Scot who, on receiving his change after paying for his tot in a pub, said, 'It's richt, mind ye, but only jist.'

13 Production

With the final technical hurdle overcome, plans for production at the David Brown works in Huddersfield went ahead fast. A sales company called Harry Ferguson Ltd. was formed and at the same time the name of the Belfast car business was changed to Harry Ferguson (Motors) Ltd. Brown had a small investment in the sales company, but Ferguson and his friends and backers, McGregor Greer and Williams, had a comfortable controlling interest. David Brown, whose father would still have nothing to do with the operation, set up a small manufacturing company called David Brown Tractors, for which he managed to raise £30,000 to pay for tooling and renting of space from the gear company. David Brown also found office space for Harry Ferguson and his staff in the old Karrier car works in Huddersfield; and he put him in touch with a Huddersfield accountant called John Turner, an inscrutable, slow-speaking Northcountryman whose manner belied a brilliant financial mind. Turner was to remain Ferguson's chief guide and adviser in all the many financial machinations inherent in big business.

The Fergusons rented a house at Honley just outside Huddersfield and made quite a stir by installing a butler. For, in addition to his excellent taste in engineering design, Harry Ferguson had excellent taste in other matters—with the exception of the sweet white wines he drank if he drank at all—and gracious living came easily to him, even if he had not been brought up in gracious surroundings.

The first pre-production tractors came out of the Brown plant in the early part of 1936, and in March Ferguson was able to give a private demonstration to Lloyd George on his Surrey estate; as a result Lloyd George ordered a tractor. The production version weighed about the same as the Black Tractor, but the engine had been changed to an 18–20 hp Coventry Climax in place of the Hercules that had powered the prototype. The range of implements to be marketed with the tractor were a two-furrow plough, the aforementioned tiller, a three-row ridger and a three-row cultivator. The price of the tractor was £224 and each implement cost £28. (The Fordson at this time was being marketed at about £140.)

The Brown-Ferguson, as it was usually called, was painted grey. Ferguson insisted on a darkish colour, but he had to admit that the black of his prototype was hardly practical if the tractor was always to appear clean. He and his team therefore compromised on battleship grey, and this remained the standard, unaltered colour of Ferguson equipment as long as its inventor had a say.

The presentation of the tractor to the public was at a demonstration near Hereford in early May 1936. As was to be expected, many farmers scoffed at their first sight of it. They were startled when a few minutes later they saw it making light work of turning two 12-inch furrows in stiff land. Then, when the farmers were already surprised, Ferguson pulled one of his tricks of showmanship: he disconnected two of the spark plug leads and the tractor still murmured easily along. At this and other demonstrations in Ulster, Eire and Derbyshire, holes were dug in the furrow bottom and planks thrown in the path of the tractor wheels to show that the plough maintained an even depth of work despite the tractor's violent pitching. The farmers' response was excellent wherever the tractor was shown.

In June 1936, the first one was sold: it was number 12 off the production line (the others having gone as demonstration models) and it was bought by John Chambers's father. He had of course been following the progress of the Ferguson System, as it was by then known, with interest since his son had been deeply involved in the design of the Black Tractor.

'My father had an old Austin tractor and six horses on the farm,' John Chambers remembers, 'but when he got his Ferguson he was able to get rid of four horses and sell the old Austin as well.'

This was exactly the sort of impact that Ferguson had always been aiming at and that which his machinery was designed to achieve. For he abhorred the use of horses as a means of obtaining tractive power on the land. Firstly, his own childhood experience with horses had no doubt soured him, but in addition he had well-reasoned and valid grounds for stating quite categorically that horses had no place on the farm, a doctrine that he preached incessantly from about 1936 onwards and which was the basis for some much wider economic ideas that he later developed. His grounds for condemning horses as a source of tractive power were that a team of horses consumed the produce off five acres of land in a year. In Britain, and particularly in the Ulster that he knew so well, the farms were generally small and the use of so much land just to supply the power needs for tilling the rest of the holding was a crippling economic handicap. And he knew that in many parts of the world farms were just as small, if not smaller; he saw his small, lightweight tractor with its integrated implements as the key to improving the level of life for the millions of smallholders in the world. Quite apart from its economy of operation, its hydraulically mounted implements ensured that not a square foot of smallholding needed to remain untilled. It was to show this that he devised, as part of his demonstrations, the setting up of a roped off enclosure with just one opening to represent a gateway. He would drive the tractor into the enclosure with a tiller mounted on the back and, with a few deft passes, cultivate the whole area, backing the tiller into the corners so that not an inch was left untilled. Finally, pirouetting the tractor into a final pass he would drive out leaving a perfectly cultivated square without a single wheel mark in it.

Ferguson believed that his tractors and implements would be able to solve Ulster's economic troubles of the thirties. Ulster's farm produce could be increased fourfold with the help of mechanisation and proper use of fertilisers. There should be an agricultural revolution to follow the Industrial Revolution. Britain was importing £400 ($2000) million-worth of food a year, and some of those imports were utterly ridiculous, he believed. For example, why import maize for pig food when home-grown barley would do just as well? And yet, despite the heavy imports of food, the area under cultivation in Britain was steadily dropping. 'Prosperity in agriculture means all round prosperity,' Harry Ferguson repeatedly said, and wrote too in articles and letters to the press. 'The smaller the farm the less it can afford to keep a pair of horses, even if farmers are as loth to part with them as the army were with their spectacular cavalry regiments.'

The plant that David Brown set up to produce the tractor at Huddersfield was in many ways a model of advanced engineering for the times. Staff from the *Implement and Machinery Review* visited it and were impressed. Micrometer controlled boring jigs were in use '. . . surely the last word in precision work,' as the *Review* commented. There was also a bolt-making plant and nuts were cyanide hardened for 'increased service and durability'. (Ferguson would never have tolerated nuts of which the hexagons might burr.) While on the topic of nuts and bolts, it is interesting to note that he allowed the use of only two basic size of hexagon head on the tractor and on all the implements—one was $\frac{11}{16}$ inch AF (across flats) and the other $1\frac{1}{16}$ inch AF. With the single spanner that was provided with the tractor, the driver was able to make all the adjustments he ever needed as well as tighten the bolts on the tractor housings from time to time. In addition, the spanner was exactly 10 inches long—the width of a normal furrow slice—and was marked off in inches so that the driver could use it as a measure when making field adjustments. The standard nut sizes and the plough spanner were in line with Ferguson's obsession for time-saving, convenience and efficiency.

Reverting to the Ferguson-Brown tractor plant, every engine was tested for its power output before leaving the factory, and the machinery was carefully rustproofed before receiving the top coat of Ferguson grey paint to 'show finish standard', in the terms of the trade. According to the *Implement and Machinery Review,* quality control was 'ruthlessly applied at all stages'.

But just as impressive as the production plant were the measures taken by Ferguson's marketing company to ensure good service and proper use of the equipment. For Ferguson must have been one of the first industrialists to realise the need for product training, as it is now generally called. One of the first measures he took when tractors began to come out of the David Brown plant was to set up a training school for farmers, operators and dealers. This school gave two-week courses in the handling, maintenance, and correct adjustment of the equipment. It operated in a one-acre field to start with, but ultimately a 50-acre farm was used. The school was run by a farmer's son, Bob Annat, who had always wanted to be an engineer and had had such rows with his father about it that he ran away from home at sixteen. He first met Ferguson at a Huddersfield demonstration; a local lawyer, Max Ramsden (later Sir Max), who wanted to see the tractor that Brown was thinking of building, had asked Annat to accompany him and give an opinion on it. Annat was so taken by the little tractor that he told Ramsden that he would like to work for Ferguson. Ramsden, who later put money into the Ferguson business at Huddersfield, passed the message on, and Ferguson called Annat for an interview at which he came under sharp scrutiny from those glinting blue eyes. At one point Ferguson said, 'You look as though you always want to be doing something with your hands, and you've got hands that look as though they can do something.'

The school that Annat ran was necessary partly because the equipment was so completely different from any which had been previously available, but also because its correct use was vital to its success. To take one example, if a disc coulter was set too deep, so that its hub was running on the ground, it would carry the weight of the plough, destroy the weight transfer and give rise to wheel spin.

Ferguson himself took a great interest in the school and often arrived at the end of the day to see how things had gone. He first cast a glance at the tractors and implements to make sure that they had been perfectly cleaned before being put under cover for the night, and then he fired questions at the pupils. If Annat intervened he was brusquely told to keep quiet, and if a question was not satisfactorily answered, or a practical exercise such as hitching an implement onto the tractor not carried out in the exactly prescribed manner, he held Annat responsible. For every single detail of the tractor and equipment's use and maintenance had been precisely laid down in order to obtain maximum efficiency and length of life from it.

Ferguson was also fanatical about service, a point which was seldom given the attention it deserved by tractor companies. A record card was kept on every tractor sold and urgent service needs received the promptest attention. The company once received a 2 a.m. telephone call from the distributor in Norway to say that something had gone wrong with his demonstration model. The same afternoon, John Chambers was aboard the Norway-bound packet as it steamed out of the Tyne, on his way to see to the matter. Such service induced the *Farm Implement and Machinery Review* to describe the company as 'an organisation of the highest integrity, anxious to intensify its good will'.

The tractors did not in fact give much trouble once some early faults had been eradicated. One of these faults was caused by a production error that made it necessary to replace the crown wheels and pinions in the first fifty tractors, an expensive operation. Another fault concerned the steel bush in which the hydraulic control valve was set. This bush was pressed into the aluminium of the pump body and when the oil warmed during work, the aluminium expanded at a different rate from the steel; the result was that the bush became loose and sometimes fell out. When this happened the tractor became entirely useless since the hydraulics were non-functional. This first happened on a tractor in Essex in 1936, and Sands was called upon to devise a solution. He designed a clip that could be fitted to the bush to give it positive axial positioning in the pump housing. All the tractors already on farms, close on a hundred by then, were visited and modified. But then there were troubles when the bush began to *rotate* in the pump housing; Sands had to modify the clip by adding a tongue that fitted in a groove on the bush and prevented any movement at all, in any direction. Once again, all the tractors were revisited and remodified. Ferguson was very irritated by this double operation. 'I thought you'd cured the trouble the first time,' he snapped at Sands.

'I thought I had too,' Sands flared back at him in a rare moment of open ire—and about the only one in which he is known to have been brutally frank with Harry Ferguson. 'I suppose a clever bugger like you would have realised that the sleeve would rotate and cause trouble even after it had been fixed axially!' Ferguson dropped the subject hastily, for he valued Sands too much to want to offend him.

Apart from the education and service offered, the company formed an aggressive marketing policy. Ferguson realised that he had a product so far ahead of its time that it might prove difficult to convince people of its worth—particularly farmers with their inherent conservatism. He therefore made statements to the press that were exaggerated to say the least. 'An extraordinary demand has already been received for the new lightweight tractor that has been placed on the market by Messrs Harry Ferguson Ltd of Huddersfield' was one of the sentences that he planted with a newspaper. But he also organised a campaign to show just what the tractor could do. Some of these demonstrations he gave himself, but later most of them were given by Knox or Annat.

Knox gave a noteworthy display at the Rowett Research Institute, Aberdeen, in January 1937. The demonstration resulted when the manager of the Institute, Arthur Crighton, ordered a Brown-Ferguson tractor at the Smithfield Show, the annual exhibit of fatstock and machinery in London. Other manufacturers at once said that, since the Rowett Research Institute was a publicly owned establishment, they should all be allowed to show what their products could do. Crighton agreed and issued an open invitation for all to appear at the Institute in January. Representatives of Ford, Allis-Chalmers, John Deere, Oliver and Massey-Harris arrived with tractors, as did Knox with the Brown-Ferguson. They prepared their equipment, but on the appointed day the rain was coming down in such torrents that even the hardy tractor experts were daunted and the demonstration was postponed. The next morning there was snow on the ground but it was decided that the demonstration would go on anyway, since a large number of farmers had arrived.

The demonstration turned into a farce: most of the tractors struggled along for a while, with much wheel spin and slippage in the snow, and all but two makes were forced to abandon their efforts to plough. One of the remaining tractors was a Massey-Harris four-wheel-drive model that was able to get sufficient traction to work, and the other was the little Ferguson, trundling along and turning two furrows that were very neat considering the conditions. This performance, while the heavy machinery was parked in baleful silence on the headland, and while the only other tractor able to work at all was several times as heavy, complicated and expensive as the Ferguson, was so convincing that Crighton immediately confirmed his order for a Ferguson; and the farmers went away amazed.

However, it was only in Scotland where the land was hilly and stony that the Brown-Ferguson made any impact on the market. The traction given by weight transfer, the safety provided by the three-point linkage, and the protection of the implement through overload release, were quickly recognised as unique features. For the same reason, Brown-Ferguson tractors were well received in Norway. But elsewhere, despite the promotional campaigns, the excellent service and the outstanding performance of the tractor itself, no one was interested in it. One of the reasons was probably that many farmers owned trailed implements for the Fordson and other tractors and they would not buy a new tractor on which these could not be used. Another reason was that the tractor was too different, too novel, and the basic scepticism caused by its light weight was difficult to overcome even when its performance was demonstrated; farmers were not noted in those days, and are not today, for their propensity for innovation. And, of course, the mid-thirties were an unfavourable period for agriculture; there was such disorder in the world's economies that farmers were unable to sell all they could have produced; at the same time, industrial workers were malnourished but did not have the purchasing power to acquire more food. There was no rush to mechanise agriculture in such difficult circumstances.

With so much militating against tractor sales, it proved difficult to dispose of the production that the Brown plant was capable of turning out, and the marketing company was in difficulty by 1937. The tractors were beginning to accumulate in stock, even though only some fifteen to twenty a week were being built. This situation inevitably caused strain between Harry Ferguson and David Brown. Brown found himself in particular difficulty because the tractors in stock were being held by the manufacturing company, and the sales company would not take them until they had orders for them. Ferguson, who was making very little money on the tractor but had a good margin on the implements, told Brown that he should cut the price of the tractor and step up production. Brown considered that the tractor required design changes before it would sell and he therefore suggested that some market research be carried out to establish exactly what those changes should be.

Ferguson refused point-blank to have anything to do with such a scheme, for any idea of changes to his brainchild were heresy to him. As Brown recalled, 'The battles we had to change even a split pin!' Obviously, therefore, the major changes that Brown felt to be necessary were utterly out of the question. Nevertheless he pushed ahead with his plans for a market study; and farmers' replies seemed to confirm his notions. They wanted a more powerful tractor that would pull implements other than those designed for the Ferguson System, and they wanted more implements than the four basic ones that Harry Ferguson Ltd. was then able to offer.

After the first year of operations, Ferguson and his family moved back to Belfast, and this made it more difficult for him and David Brown to resolve their difficulties. Ferguson wrote lengthy letters exhorting Brown to cut the price of the tractor and increase production volume, but they were at loggerheads. Finally in the early summer of 1937, the stage was reached where the sales company was in deep financial trouble. They had been spending liberally on promotion and advertising but the market was not responding. The only solution was to merge the manufacturing and selling companies; David Brown Tractors, even though making little money, was in a much better financial situation than Harry Ferguson Ltd. In June 1937 the accountants worked out the details of the merger and Ferguson-Brown Tractors Ltd. was formed. Ferguson and Brown were joint managing directors, but Ferguson and his supporters were minority shareholders, a fact that pleased them little.

'The immediate success of the Ferguson hydraulic farm machinery, coupled with the difficulty of meeting demand' was the reason the press gave for the merger—which goes to show that the company was wide awake to the politics of marketing. In point of fact, it was not an entirely dishonest approach; all the experts in farm machinery were virtually unanimous in their praise for the Ferguson equipment. One report, for example, described its 'efficiency and ease of operation effected by the automatic hydraulic control, which at first appears almost uncanny and yet in practice has proved to be astonishingly simple and reliable', and praised its 'delicate precision of work', and 'exceptional economy'. But the farmers obstinately refused to buy it in quantity.

Finally, in desperation, Brown persuaded Ferguson to come to Huddersfield to discuss the whole matter yet again. During the meeting, Brown insisted that they make a more powerful tractor. Ferguson was equally insistent that if the needs of small farmers the world over were to be met, the tractor should be smaller and lighter still. 'Look, we've got to do something about this,' Brown said in the end. 'I am going to get together some engineers and we are going to design a tractor to fit the specification that we have drawn up to satisfy the needs we have discovered during our market research.'

There was nothing that Ferguson could do to prevent this and he returned to Belfast, very upset. David Brown gathered together the engineers and draughtsmen he needed, put them in offices next to his so that he could keep an eye on them, and the work on a new tractor began.

Strangely, Ferguson did not inform his staff in Huddersfield about Brown's initiative, but gradually the news leaked out, first to Chambers who was Ferguson's engineering representative at the tractor plant. Since he knew the Brown engineering staff, it was soon obvious that some secret project was in progress. When he was able to discover that it was a new tractor and that the plan was to present it to Ferguson and manufacture it whether Ferguson liked it or not, he wrote a letter of alarm to Sands saying that he had something very important to tell him. He suggested going to Belfast, but Sands replied that he would take the boat to Heysham and they could meet there. On the following Sunday, Chambers drove to Heysham and met Sands off the boat. They spent the day together at Morecambe and Chambers told him about the tractor that was being designed by David Brown. Immediately on his return to Belfast, Sands went to Harry Ferguson's office above the motor premises in Donegall Square and broke the news—news that he thought to be momentous.

'Oh yes,' said Ferguson airily, 'I know about that.' And the subject was dismissed. Later, Knox went from Huddersfield to Belfast for a week-end and repeated the news to him, and again it was dismissed as inconsequential. The fact was that to Ferguson the matter was inconsequential, for over a period of months prior to Brown's definite decision to design another tractor, he had been laying the foundation for other plans.

14 *The Handshake Agreement*

Ferguson had never lost sight of his objective of 1920 when he had asked Ford to mass-produce his plough: he was convinced that only large volume production at low prices would enable his inventions to be marketed on such a scale as to make an impact on agriculture, and in particular on the lives of farmers. It was for this reason that he continuously exhorted Brown to reduce his price on the tractor and increase production. But Brown could not agree with this philosophy; he was much more pragmatic in his approach to business and was not carried away by great concepts as was Ferguson. Ferguson therefore began to consider Henry Ford again, the only person in his view who fully understood the implications of mass production, and was also capable of thinking in the grand manner of idealists.

After the collapse of the Ferguson–Sherman plough business (upon the withdrawal of the Detroit-built Fordson from the U.S. market in 1928), Ferguson had remained in regular touch with the Sherman brothers. They ran a company called Sherman-Shepherd which was importing British-built Fordson tractors into America, and therefore they maintained close links with the Ford Motor Company of Detroit. In addition, Eber Sherman, the elder brother, was on friendly personal terms with Henry Ford.

For this reason Ferguson kept Sherman informed of his developments in Britain and suggested, in early 1938, that Sherman come to Britain to see the Ferguson-Brown tractor at work. Sherman came, and saw a spectacular demonstration. He had always had a high regard for Ferguson's inventions, and he went away determined to draw Ford's attention to the Ferguson-Brown tractor.

Luckily, Henry Ford was experimenting with tractors again at the time. In fact, he hardly ever ceased fiddling with them, so dear to his heart was the idea of improving farming and raising agricultural production. For he, like Ferguson, considered that agriculture was the basic industry of the world, and therefore the one capable of bringing the greatest benefit to mankind if efficiently run. Prior to launching the Fordson in America in 1916, he had spent much time and money experimenting with tractors, and when the Fordson started to come off the production line he passed much of his working day in the plant, in a sixteen-feet-square glass office stolen from a corner of the building. (He always disliked paper work so he was probably happiest in a corner of the factory.) He received a British journalist in his glass office in 1916 and told him vehemently, 'I care nothing for your war. War is destruction, and all I care for is the productivity of the soil, and not least the English soil.'

In later years, during and following the depression when farmers' incomes were particularly low, Ford became a champion of the agricultural community. He considered that one way of helping would be to find industrial uses for farm produce so that, in times of surplus, prices to the farmer could be kept more or less stable. He spent large sums of money on research in this sphere and succeeded in using soya beans for making plastics. From these he actually made a tonneau cover for a car as well as gear lever knobs. He also spent considerable resources attempting to discover a plant that could yield latex on a commercial scale.

In the mid-1930s, owing to his declared intention of doing something to help the farmer, he intensified his efforts to produce a tractor that would be an improvement on the Fordson. By the summer of 1938, he had an engineer called Karl Schultz working for him on a special assignment to design the new tractor. Schultz built three prototypes; the most promising, but still far from being satisfactory, was a three-wheel version with a V8 engine, the drive being taken from a single rear wheel. (Ferguson once said that three-wheeled tractors were 'engineering abortions' and he had a point.)

To assist with these field evaluations on the Ford farms at Dearborn, a young man called Ed Malvitz was employed as tractor driver and assistant to Schultz. Some of the tractors were very rudimentary and Malvitz remembers that at least one of them was a 'cobbled up design . . . that was practically no good'.

Henry Ford himself took a deep personal interest and went to the field almost daily to question Malvitz on the performance of the various designs.

It was in this atmosphere, with Ford trying to fulfil his promise to help the farming community, that Eber Sherman arrived back in North America having witnessed the demonstration of the Ferguson-Brown. He informed Ford of the remarkable system that Ferguson had invented and suggested that it could be of interest to Ford. Ford, remembering the occasion about eighteen years earlier when Ferguson had demonstrated his plough at Dearborn, asked Sherman whether Ferguson would come to Dearborn again. This, of course, was exactly what they had been angling for. Accordingly, in the autumn of 1938, Ferguson prepared to go to the United States to show Ford his tractor. He arranged for a tractor, number 722, and a set of implements to be sent to Belfast, from where they were shipped to America. He told David Brown that he was going to America, but he did not tell him why, vaguely dismissing the question when Brown asked him whether he would be seeing Ford.

In October 1938, Harry Ferguson and John Williams sailed to New York where they met the Sherman brothers and arranged to truck the crated tractor and implements to Dearborn.

Henry Ford was visiting some of his mines in Northern Michigan when Ferguson, Williams and the Sherman brothers arrived and he did not return for several days. The time was spent uncrating the tractor and implements, assembling them, and selecting the most suitable place for the demonstration; Ferguson wanted a field near Fairlane, Ford's private residence, and one that was free draining in case it rained shortly before the demonstration. He finally settled on a small field close to the tree-lined drive that leads up to Fairlane, a slightly inclined clearing that Mrs Ford used as a plant and shrub nursery for the main gardens that surrounded the fine house. The tractor was tried and prepared down to the last detail. Ed Malvitz was amazed by its performance and subsequently told Henry Ford that it looked to him 'as being ten to twelve years ahead of the competition'.

The day of the presentation was a classical North American October day, fresh and slightly misty in the morning with bright, warm sunshine in the afternoon. At three in the afternoon, the group gathered in Mrs Ford's nursery garden. It was a small group, for only Henry Ford with his chauffeur Wilson, Ferguson, Williams, George and Eber Sherman, Ed Malvitz and the

Sherman-Shepherd employee who had driven the truck and helped assemble the equipment, were present. Ferguson had arranged for Malvitz to drive the tractor so that he could stand next to Ford and give him a running commentary.

The tractor ploughed a few rounds, then Ed Malvitz changed the plough for the tiller and carried out the usual Ferguson demonstration of cultivating in a tiny roped-off enclosure. He followed this by showing the work of the ridger and the three-row cultivator, changing implements with a facility that in itself was impressive.

Legend has it that the tractor had barely ploughed a couple of rounds before Ford rushed over calling 'Stop, stop! You've got it; this is brilliant,' or words to that effect. Malvitz, the last witness of that demonstration, cannot remember any such happening, however. He remembers that both Ford and Ferguson took on a serious air. Ford seemed to be giving the tractor and its performance some very thoughtful consideration, and this seems to be borne out by the actual events of that historic afternoon.

For hardly had the demonstration with the Ferguson-Brown finished than Ford asked that an Allis-Chalmers model B tractor and a Fordson, from the estate's pool of machinery, be brought to the nursery for comparison. Both of these tractors suffered from considerable wheel slip as they ploughed up the gradient with its loose surface, whereas the Ferguson-Brown went up, using the same size of plough, without hesitation.

'Well,' said Henry Ford after that, 'you're ploughing in my wife's garden here. Let's see how the tractor does in some heavier going. We'll take it into the Deer Field.'

'I'm glad you suggested that,' replied Ferguson, 'because that is something I'd like to show you too.'

The Deer Field, of over 250 acres—the largest field on the Ford farms—lies just on the other side of the drive from the nursery plot, which today is overgrown. (Michigan University, which uses Fairlane, is obviously less interested in gardening than was Clara Ford.) The Ferguson-Brown was put to work in the heavier soil of the Deer Field and again it excelled itself. Ford wandered over to one of the old and untended apple trees in the field, picked an apple, pulled out his pocket knife to cut out the wormy bits, and munched it thoughtfully. Then he called his chauffeur Wilson and told him to fetch a table and two chairs from the nearby lodge at the entrance to Fairlane. These were fetched and set up in the field; Ford and Ferguson sat down to look at a spring-driven model of the tractor that Ferguson had brought and which illustrated the advantages of his three-point linkage; and they began to talk business.

The two men started out from common ground: both had been brought up on farms and were deeply interested in seeing agriculture and farmers gain the recognition and reward due to them. They were especially distressed over the hard time that farmers in both America and Britain were having in the thirties. In addition, Ferguson was able to convince Ford that his tractor was suitable for use on even the smallest farms, thus making it possible to eliminate draught-animals that produce tractive power at such a high cost in terms of acreage needed for their support. Wholesale farm mechanisation, he persuaded Ford, would bring undreamt-of benefits to farmers and to all mankind, and establish agriculture in its rightful place as the number one industry.

This ideology sparked an enthusiastic reaction from Ford; the two men had an extraordinary affinity of idealistic thought. It is interesting to note that there was even a certain physical similarity between them, for apart from their common attributes of being original, strong-willed and eccentric, they both had a beaky, aesthetic appearance that has caused some people to comment, when looking at photographs, that they looked like father and son.

Having firmly established, as they sat at that table in the field, that they shared many opinions on the subject of agriculture and its mechanisation, they began to discuss ways in which they could work together. Firstly, Henry Ford offered to buy the Ferguson patents for a large sum. 'You haven't got enough money, because they are not for sale at any price, to you or anybody else,' Ferguson told him bluntly.

'Well, I'm determined to go into the business, and you need me and I need you. So what do you suggest?' asked Ford.

'A gentlemen's agreement.'

'What do you mean?'

'Well,' explained Ferguson, 'you're proposing to stake your reputation and your resources on this economic idea, even if a billion dollars is involved, and no agreement could protect you fully. I've spent my whole career thinking out this great economic idea and I've put everything I and my family have into it and I reckon my designs and inventions are worth more than your billion dollars, so I don't see how I can make an agreement. I'll trust you, if you'll trust me, and I'll put my services at your disposal for future designing, education and distribution. It'll have to be world education because there's a new world economy involved. For your part, you'll put all your resources, energy, fame and reputation behind the equipment and manufacture it in volume, at low cost. I'll sell it.'

'That's a good idea,' said Ford, 'I'll go along with you on those terms.'

They then went on to work out a few more details and in the end came to an agreement which included five points:

1. Ferguson would be responsible for all design and engineering matters and would have full authority in this respect;

2. Ford would manufacture the tractor and assume all the risks involved in manufacture;

3. Ferguson would distribute the tractors, which Ford would deliver to him F.O.B. for sale wherever and however he pleased;

4. Either party could terminate the arrangement at any time without obligation to the other, for any reason whatever, even if it was only 'because he didn't like the colour of his hair';

5. The Ford tractor plant in Britain (which was by then at Dagenham) would ultimately build the Ferguson System tractor on similar terms to those established for Dearborn.

Obviously such an arrangement was of major industrial significance and involved many, many millions of dollars; yet when the two men sitting at that little table in the Deer Field of Fairlane estate had reached this accord, they stood up, shook hands warmly and decided that the agreement was one of confidence and trust between gentlemen. As such, the clasping of hands was sufficient and no formal written agreement needed to be drawn up. This famous 'handshake agreement', as it has since become known, is almost certainly unique in the world of big business. There was never a scratch of pen or pencil to record it at the time. The version given here is how Ferguson remembered it.

In the days following the handshake agreement, the Ferguson-Brown tractor was demonstrated to most of the Ford top management, including firstly to Ford's son Edsel. All were impressed, but a number of people were much against such an important arrangement being agreed and sealed by a mere handshake. However, Ferguson and Ford, kindred spirits, wanted it that way, and that way it remained.

Mightily pleased with the way events had worked out, Ferguson and Williams returned to Britain. The first thing they would have to do was to withdraw from their partnership with David Brown, on the most favourable terms possible. Ferguson realised that Brown's initiative in designing a new tractor model could be very useful to him if he played his cards right. And playing them right meant first of all playing them close to his chest.

When Brown met Ferguson after his return from the United States he felt that it would have been a very natural thing for Ferguson to have renewed his old contact with Ford, so he asked him outright whether he had seen him.

It was too soon for Ferguson to show his hand. 'Oh, you can't get near that man,' he replied.

It was incidents such as these that have caused some Ferguson critics to describe him as 'devious', and in a sense he certainly was. On the other hand, almost every man involved in business, big or small but especially big, has to be devious from time to time if he is to progress or indeed survive.

The opportunity to make the break with Brown soon arose. Brown showed his partners the design for the new, heavier and more powerful tractor that he intended to produce and they immediately termed this a breach of the original agreement that had launched the Ferguson-Brown co-operation. They therefore caused a row and made it known that they wished to withdraw from the company. David Brown was pleased, for as we have seen, relationships had been difficult for some time. He offered to buy out their original stake in the enterprise, which they accepted as a solution. There were still a number of Ferguson-Brown tractors in stock and Brown undertook to market these and to honour any Ferguson obligations in terms of after-sales service.

The collaboration between Harry Ferguson and David Brown had produced and sold about 1200 tractors in the period between the summer of 1936 and January 1939, when the break was announced. In July 1939, Brown launched his own tractor with a hydraulic lift. Today, David Brown tractors are among the world's best; it was through that brief and often stormy association with Ferguson that the company entered the tractor industry in the first place.

15 The Partnership at work

In January 1939, the adventure with Henry Ford began. Harry Ferguson, with his wife Maureen and daughter Betty, set sail for America. The team that he chose to take with him consisted of Willie Sands, John Chambers, with their families, and Harold Willey who had been on the sales staff of the Ferguson-Brown company. They sailed aboard the *Aquitania* and arrived in New York on January 14th.

Ferguson, his wife and daughter set up quarters at the Dearborn Inn, a hotel which Henry Ford had built in conjunction with the airport that he had established at Dearborn in 1924. Incidentally, this airport, which was closed in 1933 and is now the Ford Motor Company's proving ground, was the base for the first scheduled passenger service in the United States; and the first all-metal, multi-engine passenger-carrying aircraft in America was built there. In addition, the Ford airport was the first in the world to guide a commercial airliner by radio, and the Dearborn Inn was the world's first airport hotel. The Inn, a comfortable colonial style building set among trees, has a number of small houses close by which it leases. The Ferguson family moved into one of these—the Patrick Henry house—a pleasant lapboarded building just behind the Dearborn Inn. They had been in the house only a few minutes and were still getting their bearings when the lean and sprightly figure of Henry Ford walked in, for he had come to welcome them in person. Harry Ferguson immediately told Betty to make a cup of tea for them all, and she disappeared into the kitchen to do as bidden. To her horror, she could not get the electric cooker to heat, and she looked in vain for the mains switch. Knowing her father's intolerance of failure, she hardly dared go back to the sitting room to admit that she could not make the tea. But finally she returned to the others.

'Mr Ford,' she said, 'you don't happen to know how to turn on the electricity, do you?'

To her relief the question provoked laughter and they all went to the kitchen to see if they could solve the problem but it was too much even for the combined ingenuity of those engineering pioneers.

During Harry Ferguson's absence in England, when he had gone to settle matters with David Brown and collect his team before returning to Dearborn, the Ford Motor Company had not been idle with the tractor. Firstly, they took the Ferguson-Brown that had been used for the demonstration to the old Ford airport building and completely stripped and reassembled the transmission and hydraulic system. They wanted to see how it worked and also determine what modifications would be necessary for Ford-type mass production. They also carried out a series of laboratory tests and were astonished by a number of things, but especially by the fact that the greater the draft load on the implement the more the weight on the front wheels of the tractor. People were still used to a lightening of the front of the tractor as the draft load increased: that the compression load in the top link exerted its forward and downward pressure through the tractor and thus stabilised the front wheels was still a novel concept for those who had not followed the Ferguson plough developments closely.

Apart from these examinations and tests of the Huddersfield-built Ferguson-Brown, Ford had also constructed three prototype tractors incorporating the Ferguson System during November and December 1938 and sent them to his Richmond Hill plantation at Ways, Georgia, for test. (Field testing of agricultural machinery in Michigan and nearby states is impossible during the winter owing to frost and snow.)

When Harry Ferguson and his team arrived back in Dearborn in January 1939, the first task was to finalise the production design of the new tractor. The Ford and Ferguson engineers worked closely together, firstly in the old airport terminal building but subsequently in the so-called Blue Room above the car experimental workshop in the Rouge plant. The Huddersfield-built tractor was taken there, and so was the complete set of blueprints and drawings the Ferguson team had brought with them from Britain.

Some of these drawings covered improvements on the Ferguson-Brown tractor, among them a simple but clever front axle that Sands and Greer had worked out: in order to be able to adjust the front-wheel track for row crop work in the easiest way possible, the new axle was to be made up of three beams; the two outer beams bearing the wheels were bolted to a central beam, and the whole was so laid out that they formed a part of an arc or semicircle. Sliding the wheel-carrying beams of the axle in or out to alter the wheel track carried them in an arc so that the radius rods and the steering linkage did not need to be altered. This idea was used later by a number of other manufacturers; it was never patented owing to shortage of funds at the time of its invention.

The Ferguson team also took with them to Dearborn the design for a power-take-off shaft that the Huddersfield-built tractor had not incorporated. The power-take-off, a shaft protruding from the rear of the tractor to which implements such as grass mowers could be coupled to obtain their drive, was already essential for farming conditions of the late 1930s. Indeed, the lack of a power-take-off had been one of the things that David Brown complained of in the design of the Ferguson-Brown. So strongly had Brown felt on the subject that he had had his own engineers contrive a power-take-off shaft to be fitted to the Ferguson-Brown tractor as an afterthought. Inevitably, such an arrangement was a botched-up contraption and Willie Sands, who was irate about it at the time, was still so thirty years later. The power-take-off design they took to Dearborn was one that they had drawn in 1937; it was properly engineered and the shaft protruded rearward from a position set within the triangle described by the attachment points of the two lower and one upper links. This was the most convenient means of being able to drive mounted mowers, potato diggers, and similar implements. The arrangement was patented in 1938.

Henry Ford had promised to put the whole staff of the Ford Motor Company, including Charlie Sorenson and all the company's resources in factory and field, at the beck and call of the tractor programme, and he was true to his word. The work went ahead at an amazing rate; Sorenson was in the Blue Room (and in the other offices to which by now the work had spread) day after day to urge things along. Harry Ferguson, who insisted that the designs should be submitted for his approval at every stage, also gadded from one office to another. 'Not a bolt or a nut could be put in there without his consent,' was one subsequent comment. In general, however, it was a question of making the tractor designs more suitable for the Ford mass-production facilities, and so long as the issues at stake were only these, there were few arguments. But Ferguson, Sands and Chambers were as stubborn as Ulstermen arguing over religion when a more fundamental issue was being discussed. At one stage, when Ford wanted to fit an engine from one of their car models with very little modification, Ferguson insisted adamantly on major changes to—among other things—the cooling system, and the flywheel. He also took the opportunity of unblushingly giving Ford engineers some of the benefits of his knowledge of engine design; for example he told them, rightly, that overhead-valve engines were better than side-valve engines.

Friction arose, but the work was not seriously hampered. It is remarkable that by the end of March a new prototype was ready—well under three months from when work had begun in the Blue Room. The prototype was hardly pretty because the styling was not finalised and it had a high and ungainly bonnet or hood, but it incorporated all the ideas and inventions of the Ferguson-Brown plus the others that the Ferguson team had brought with them to Dearborn.

Henry Ford and Harry Ferguson decided that the completion of the prototype should be marked by a suitable 'occasion' and on April 1st, 1939, it was taken to Mrs Ford's nursery garden close to Fairlane where a few carefully-chosen guests were invited to see it work. The Ford and Ferguson families were there, as were also the Sherman brothers and a small number of people who were already interested in marketing the new tractor. Speeches were made in which it was stressed that this was 'an historic occasion' and then Ferguson climbed onto the tractor and ploughed a length of the plot; at the end he got down and Henry Ford took his place to plough the return leg, thus ensuring that both members of the unique partnership were making history together. It is said, though it is impossible to verify, that Ford on his return run ploughed in some of Mrs Ford's raspberry canes; historic occasion or not she told him quite clearly what she thought of his performance.

Following the demonstration, Henry Ford gave orders that the 'historic furrows' were to be fenced off; he almost certainly had in mind preserving them for posterity in his Greenfield Village Museum.

During the period of technical and engineering work, Ferguson and the Sherman brothers planned and arranged the marketing of the tractor and the supply of implements to go with it. Their first step was to establish a company called the Ferguson-Sherman Manufacturing Corporation. Naturally, the expenses of getting the Ferguson team to America, of establishing the Ferguson-Sherman Manufacturing Corporation, of making contracts with implement manufacturers and of setting up a skeleton distribution and sales network were considerable, but Henry Ford generously helped by lending the Ferguson-Sherman Corporation some $50,000 to tide it over until it could begin trading. Meanwhile, tooling up for production of the tractor was going ahead in the B building of the Rouge plant, the building in which Eagle boats had been built during World War I. The styling of the tractor, in which both Henry Ford and Harry Ferguson had taken great personal interest, had been finalised with the help of plaster models.

There was one major point of contention between Ford and Ferguson at about this time in connection with the name to be displayed on the tractor. The details of that particular argument are not known, though Ferguson later said that Ford had insisted on the Ferguson name appearing on the tractor, even though he, Ferguson, had 'no ambitions in that direction' and even 'refused' to have his name on it. This appears so out of character that it seems more likely that there was a fight as to *how* the name should appear, and Ferguson, not wanting to be relegated to some tiny plaque (perhaps incorporated in the patent acknowledgement plate) said that if that was to be his only mention he did not want one at all. In any case 'having almost had a tiff about it', in Ferguson's words, it was finally agreed that the tractor would be known as a Ford tractor, but that it would prominently carry a plate inscribed with the words 'Ferguson System'. 'Ford tractor with Ferguson System' was therefore the proper name for the product of that joint venture, but the tractors have always been known more simply as Ford/Fergusons.

In May 1939, the press of North America became curious about the forthcoming tractor and the strange deal that Henry Ford had made with this Ulsterman, who was unknown in America, except to those with a special interest in farm machinery. In particular, the relationship between Ford and Ferguson—which later caused *Fortune* to run a long article under the title of 'Henry Ford's only Partner'—was intriguing to outsiders. When pressmen in Detroit attempted to establish the facts by asking the 'cocky Irishman' whether he was a Ford employee, or if not what his status was, his stock answer was, 'Just say I'm working with Henry Ford.'

Ford was lavish in his praise of Ferguson. He made repeated reference to his genius and predicted that he would become as 'famous as Edison Bell, and the Wright brothers'—comments that were widely quoted by the press. Henry Ford's ranking of Ferguson alongside such inventors, especially Edison, needs some elaboration if its real signficance is to be understood and if Ford's true regard for him is to be appreciated. Ford had what amounted to a hero-worship of Thomas Edison, probably the most prolific inventor of all time, and a tour of Ford's fascinating Greenfield Village Museum testifies to this. Greenfield Village, established in the late 1920s, consists of almost a hundred historical buildings moved piecemeal from many parts of North America, and among them is Thomas Edison's complete Menlo Park laboratory. Ford wanted this complex to be fully authentic, so he also moved seven wagon loads of the original soil from Menlo Park on which to reconstruct the buildings. Even Edison's rubbish hole was important, for had not Edison himself once said that what a good inventor needed was 'imagination and a scrap-heap'? The rubbish hole was therefore redug at Greenfield Village in exactly the same position relative to the buildings as it had been at Menlo Park.

But Ford's veneration for Edison was best shown at the dedication of Greenfield Village, an occasion made to coincide with the Jubilee of Edison's breakthrough in inventing the incandescent lamp. Ford invited Edison, then past eighty, to be the guest of honour; but Edison was honoured to an extent that not even he had expected. Henry Ford remembered that in his youth Edison had been a trainboy, a cigar, candy and newspaper vendor, on the Grand Trunk Railway. He had had a small laboratory and printing press in the baggage-car, until some of his phosphorus spilt and set fire to the car. As a result, Edison was summarily bounced off the train at Smith's Creek, Michigan. Ford bought Smith's Creek station building and some lengths of rail and re-established them at Greenfield Village in 1929. When Edison arrived, he alighted at the same station where, some sixty-seven years earlier, he had landed with less dignity.

Edison was made to sit down in the re-creation of his old laboratory, and Henry Ford, who had spent about $3 million just on his collection of Edisonia, asked his idol what he thought of it.

'You've got it about ninety-nine and one half per cent perfect,' said the old man.

'What is the matter with the other one half per cent?' asked Ford.

'Well, we never kept it as clean as this,' drawled Edison

When Edison got up from the wooden chair, in which he had been sitting when he made that comment, Ford sent for a worker to nail its legs to the floor, from which position it has not been moved to this day.

This story will give some idea of the significance of Ford's praise when he compared Ferguson to the outstanding inventor Thomas Edison.

Journalists' requests for interviews became more numerous still as the production line for the new tractor was being finished. Henry Ford was accompanied by one pressman during a tour of inspection in the B building of the Rouge plant and Ford, when led onto the topic of the old Fordson tractor, referred openly to its fault of rearing up when under load. He also said that he hoped it would be possible to produce the new Ferguson System tractor at a price lower than the lowest price ($495) at which the Fordson had been sold in the last months before going out of production.

By June 1939, the first of the production tractors were ready, and on the twelfth of the month the Ferguson-Sherman Corporation held a lunch and meeting at the Dearborn Inn, followed by a demonstration for the distributors who had already been appointed. This was followed on the twenty-ninth of the month by the public launching to which 500 people from 30 American States and 18 foreign countries were invited.

For the occasion a brochure called 'Invitation to the Land' was produced. Some of its wording ran as follows:

'The land! That is where our roots are. There is the basis of our physical life. The farther we get from the land, the greater our insecurity. From the land comes everything that supports life, everything we use for the service of physical life. The land has not collapsed or shrunk either in expanse or productivity. It is there waiting to honour all the labour we are willing to invest in it, and able to tide us across any local dislocation of economic conditions. No unemployment insurance can be compared to an alliance between man and a plot of land. . . .

'This is a memorable day, for we are privileged to introduce the most far-reaching series of developments in the long history of agriculture—a new system of farm mechanisation consisting of a phenomenally light-weight Ford tractor of unique design incorporating the Ferguson System of hydraulic control for a line of light-weight indestructible implements.

'The door is open now for the next great industrial movement —the decentralisation of sorely congested manufacturing centres and a return to the land for millions of farmers who should never have left it. The migration of industry to rural and suburban areas will bring economic stabilisation to the nation and to the individual. It will be easily possible—using this new machinery which a twelve-year-old boy can safely operate—for a worker to combine industry and agriculture. He can work in industry and run a small mechanised farm as well.'

The grandiose words of that statement were in part true, but at the same time a great deal of it was an oversimplification. Where is the man who, working in industry and earning a good wage, wants to run a smallholding as well? Once people have moved from the land to the cities, with whatever loss of contentment may have resulted, what is going to make them give up the distractions of the city for the simpler country life? The drift from the land is and presumably always will be—barring some cataclysmic event— a one-way flow. In addition, however wonderful the Ferguson-designed machinery, the drudgery of animal husbandry with its heavy demands on time, even in the middle of the night, was in no way eased by the Ferguson inventions; and for a smallholder to do nothing but crop raising would hardly be worth while unless his system was so intensive that in effect he was running a market garden. But if his system was that intensive, it would also have a high labour requirement because there are many garden operations that cannot be mechanised. Thus the notion of a return to the land was hardly realisable. Nevertheless, the events of June 29, 1939, at Dearborn were impressive, and the demonstration and the Ford/Ferguson partnership got full press coverage.

'Newest addition to the royal family of the Ford Motor Company is hawk-beaked, cocky Harry Ferguson of Belfast, Ireland' was the lead sentence in the *Detroit Evening Times*'s piece the next day. It went on to report the speech that Harry Ferguson—'5 foot 6 inches, 130 pounds'—had made at the luncheon. 'This is an historical day. The recovery of industry can only be accomplished through agriculture. More than 6,800,000 families (in the United States) live on farms. Make these people contented and independent and national spending power will increase to end depressions.

'Farming is not prosperous today because it is hopelessly out of date. Six million farmers are producing crops as we did a hundred years ago. If Mr Ford tried to run his plant with power from horses he would be broke within a year. This tractor was not built to compete with other tractors—it was built to compete with horses.'

On the basis of this statement the *Chicago Journal of Commerce* opened its story on the new tractor by saying, 'It was a bad day for old Dobbin at a Ford tractor demonstration at Dearborn yesterday.'

Incredible as it may seem now, in 1939 there were in fact still about 17 million horses and mules at work on North American farms, so Ferguson's comments were not wide of the mark.

At the demonstration itself, carefully planned as always, Harry Ferguson arranged for a tow-haired boy aged eight, David McClaren from the Greenfield Village School, to plough a round in order to show the simplicity and ease of handling of the tractor. Henry Ford and his son Edsel, seated in the first row of the tented grandstand, chuckled at the sight.

Indeed, Henry Ford was ebullient about the whole occasion. 'You know,' he told one journalist, his bronzed face lighting up when he looked at the new tractor, 'this will change the whole face of farming. I don't care if a man has nothing more than a field of weeds—he can make it pay.'

Ford usually tended towards modesty with the press, but on this occasion he told them that he was not attempting to make a profit on the tractor, and added, 'I'm going in to the limit to help my country!'

Time magazine devoted a good deal of space to the new tractor. 'Scarring the green breast of one of the fields on Motormaker Henry Ford's Fairlane estate near Detroit is a 60-foot plowed furrow,' they wrote. 'Around it Ford workmen have built a fence. Over it they have laid a tarpaulin. Why this has been done no Ford employee knows for sure, but most could hazard a sound guess: it will be included in the intriguing mass of memorabilia which includes Luther Burbank's shovel (thrust into a block of concrete), and a reproduction of a hole in the ground at Menlo Park N.J. where Thomas Edison and his helpers threw their laboratory junk.' The *Time* article then gave an accurate description of the special features of the Ferguson System. Indeed the reporter was more accurate in this than in his surmise about the preservation of the furrows; for, as on the previous occasion, that of the April 1st 'historic furrows', Ford never got around to moving them to Greenfield Village. Perhaps the problem of moving ploughed furrows from one place to another, without damaging them beyond recognition as furrows, daunted even his ingenuity and means.

The technical press of the United States made very favourable comment on the new tractor and its inventor. 'Make no mistake! The new Ford performs' were headlines in the *Farm Implement News* of July 13th, 1939. 'The demonstration of the new Ford tractor June 29 before hundreds of newspapermen was a triumph for Harry Ferguson of Belfast, more lately of Evansville, Huddersfield (England) and Dearborn, for it put the seal of recognition on a man who has been telling the same story to all and sundry, the wide world over, and they largely looked and passed on. And he has been doing this for about 20 years! Now comes the reward.'

The article then explained how surprising it was that no one had adopted Ferguson's ideas on a large scale years before. The journalist had a copy of a Ferguson-Sherman plough brochure of 1926 and he found even that implement remarkable. He could hardly understand how major manufacturers had passed it over. The writer stated that at the demonstration of the Ford with Ferguson System, the tractor performed in 'such a way to make those who came to scoff leave with ungrudging admiration for its demonstrated capacities'. And he added, 'The new Ford is as quiet as a dragonfly over a sun-glint brass bar!'

The price of the tractor was also impressive: even if Henry Ford's desire to have it on the market at less than the lowest cost of the old Fordson was not possible, the new tractor price of $585 was still very much cheaper than that of other farm tractors giving similar performance on the market in North America. In some cases, the Ford price was only a little over half that of others.

Following on this success in America, the culmination of so many arduous years, Ferguson decided that he would go back to Britain to try to arrange for the Ferguson System tractor to be manufactured there. Undoubtedly he considered it a patriotic duty to have his tractor available to British farmers during the war that was so obviously on the horizon. For he had seen the effects on Britain when a U-boat blockade was slowly strangling the country, and it will be remembered that it was in assisting to increase Ireland's agricultural output in 1916–17 that he first became involved with farm machinery.

By September 1939, with the Sherman brothers looking after the affairs of the Ferguson-Sherman Corporation, he left Dearborn for New York, where he wanted to stop off at the World Fair at which the new tractor was giving daily ploughing demonstrations. On the opening day, Edsel Ford cut the first furrow; then the World Fair president, Grover Whalen, ploughed a round, and finally so did Ferguson.

Henry Ford was visiting his plants in Northern Michigan and sent the following cable: '*Regret that I do not return from the north in time to be with you at the World's Fair tractor demonstrations. I trust that a great many people will be able to see what the tractor means for the future of farming. I believe that it is the most revolutionary step that mechanised farming has taken. I look to it to accomplish two things—tnrn the farm deficit into profits and reduce the expense of going on the land. My regards to Ferguson and best wishes for a pleasant voyage.*'

16 *A Matter of Class and Personalities*

Harry Ferguson had arranged for a Ford/Ferguson with implements to be shipped to Britain before he left America, and on his arrival he gave a demonstration at the Greenmount Agricultural College, Muckamore, not far from Belfast. Some seventy officials attended, among them Sir Basil Brooke, with whom Ferguson had been in touch ever since the days of his first hydraulic prototypes and when he had hoped to establish a tractor industry in Ulster.

Already at the time of the Muckamore demonstration on October 12th, 1939, tillage orders had been sent out by the British Ministry of Agriculture. The order for Ulster was that 250,000 acres must be broken, and there was grave concern that a shortage of tractors would hamper the work. Sir Basil Brooke reckoned Ulster needed 1,400 tractors, but there were only ninety available at the time. The Muckamore demonstration was followed by several others, one at the West of Scotland College of Agriculture near Ayr. There the tractor whisked up and down a one-in-five slope turning in 100-year-old pasture. Farmers were amazed at how well it ploughed, even though 'it looked almost too frail for the job'. There, as everywhere else it went, the tractor caused astonishment and admiration. People were dumbfounded by the excellence of the work it did.

It may seem strange that Harry Ferguson bothered to give these demonstrations, but they were the tactics of a strategy that he had worked out to combat a difficult situation in which he found himself. The fact was that even though Henry Ford had agreed with him that the Ford Motor Company of Britain's tractor plant at Dagenham should also build the tractor on a similar basis to that at the Rouge plant, the management of the British Ford Motor Company had serious reservations about doing so. And they had sufficient autonomy from the parent company to be able to do as they thought best. Ferguson's demonstrations, which followed on some rather negative contacts with the management of the Ford

Motor Company of Britain, were aimed at arousing an enthusiastic following for his tractor and implements, one that he could exploit in bringing pressure to bear in the right quarters.

On January 10th, 1940, a demonstration of different makes of tractor took place near Stormont, the seat of the Northern Ireland Parliament. It is not known who was responsible for instigating this particular demonstration, but it is not difficult to make a guess. Sir Basil Brooke attended with the Minister of Commerce, J. Milne Barbour. There were many different makes of tractor present and the officials were particularly interested to see how the various types would compare, for by now it was evident that an extensive ploughing campaign was going to be necessary throughout Britain if the country was not to be starved out by a blockade. And the War Agricultural Executive Committee was going to need good machinery. The Ford/Ferguson, as usual immaculate and neat among the other much bigger machinery, made an excellent impression. It made a similar impression a few weeks later in a demonstration to the Irish Ministry of Agriculture before which Harry Ferguson gave a lunch at the Royal Hibernian Hotel in Dublin.

'History is going to be made when we demonstrate this tractor,' he brazenly asserted in his speech, but most of the people agreed subsequently that the remark had not been an overstatement.

These demonstrations and the strategy to force the Ford Motor Company of Britain to drop its trusty but unsophisticated old Fordson tractor and put the Ferguson System tractor into production in its place were to culminate in a presentation of the tractor to Ford Motor Company and British government officials in early May 1940. With the assistance of Trevor Knox, who it will be remembered had remained in Britain when most of the Ferguson team went to Detroit, a large demonstration was organised at St Stephens, Bedford. Lord Perry, the then Chairman of the Ford Motor Company, attended with many senior government officials.

During the final preparations for the demonstration, the quiet was continuously being disrupted by aircraft of the Royal Air Force passing overhead on their way to assist the embattled troops retreating towards Dunkirk. Ferguson was so irritated that he beckoned to a Ford public relations man who was in attendance.

'I can't give a demonstration with this noise going on,' he said, 'get onto the Air Ministry and ask them to reroute their aeroplanes away from here.'

The astonished Ford executive began to protest that such a thing could not be done, but he was cut short.

'Tell them it's a matter of national importance,' said Ferguson. 'It's just as important as fighting the war.'

The public relations man went away to do as bidden, but he met with no success.

At the demonstration, Ferguson stood at a table set in front of a covered stand. 'A million man-hours a day are wasted tending horses,' he declared, 'and an equal amount of man time is wasted growing food for them. Those two million man-hours could be reduced to so many minutes by the provision of 250,000 tractors and sets of implements at a cost of £48 million in capital outlay plus £3¼ million per annum for fuel. In the second year, the food produced would have a value of £10 million and in that way about 325 ships, each of 10,000 tons carrying capacity, could be relieved for other duties. We must put eight million more acres under the plough and the Ford tractor with Ferguson System can carry out this tremendous task better than any other because it can work anywhere and in any conditions. Eighty thousand units will be necessary but each tractor will only require half the steel and other materials needed for a larger tractor. Whether we can go into production depends on the government making available the material.'

The demonstration itself left most of the audience convinced that they had witnessed something of real significance, both to agriculture and the nation; but a few members of the audience, even if they were convinced, were neither willing nor able to promote the production of the Ferguson System tractor at Ford's Dagenham plant: those individuals were the very persons upon whom this production depended—Ford directors and government officials.

At least part of the resistance of the Ford directors went back to the previous year when it had seemed definite that the tractor would be produced at Dagenham. At that time Harry Ferguson began to believe that he should have a position on the board of directors of the Ford Motor Company of Britain. Neither Henry Ford nor his henchman Sorenson took exception to the idea, but the British directors did. There can be no doubt that the traditional type of British director of the period, with his upper-middle-class background, would have found Ferguson objectionable, especially if he were involved in one of his promotional campaigns in which he would harry and wrestle and try with unbelievable tenacity to outflank the opposition. His assertiveness, his outspoken claims, and his glaring intolerance of those who could not share his views, caused the Ford directors to regard him with discomfort and dislike. To them he was a presumptuous upstart off the Irish bogs.

As Lord Illingworth wrote to Lord Perry, his fellow director, in December 1939:

'I am decidedly against Ferguson joining the Board. He would be an infernal pest and no one could do any good to get a new machine going during the war, when there is an efficient one on the market already. It has been explained to him why it is impossible during the war and neither he nor Henry Ford, nor Jesus Christ can alter it.'

It would have been very difficult to make a production switch from the Fordson to the Ferguson System tractor, owing to the new plant required, even if the Ferguson System tractor and implements would have needed less material per unit once in production. Apart from this, however, the tractor situation in England at the outbreak of war was very curious indeed following an initiative taken by Patrick Hennessey, later Sir Patrick, the general manager of the Ford Motor Company in Britain at the time. About a year before the outbreak of war, Hennessey had become convinced that hostilities were inevitable and he had remembered Britain's difficulties in World War I when the shortage of home-produced food had been so critical. He was aware that such a situation would almost certainly arise again unless pre-emptive measures were taken; for once war had actually been declared, defensive and offensive weapons would take priority over something as apparently mundane as farm machinery. Hennessey therefore evolved a plan—one so commercially audacious that to persuade his superiors in the Ford Motor Company to accept it was in itself a noteworthy feat—and went to see the Permanent Secretary at the Ministry of Agriculture, Sir Donald Ferguson.

'You know, you ought to give us a contract for tractors so that we can raise our production to about a hundred a day for next year when we'll be at war,' Hennessey told him.

'How do you know?' the Permanent Secretary asked, slightly surprised.

'It's my assumption, and if you do as I suggest, when war breaks out you'll have full production of tractors and we'll store them all around the country for you with the Fordson dealers. They'll carry and look after the stock. And if there isn't a war, we'll take them back, even though that would probably mean closing down our line for about a year.'

The Permanent Secretary protested that such a contract was not possible because it would require the passage of a Bill in Parliament, but Hennessey politely informed him that he would go to the Minister himself, or the Prime Minister if necessary, if the suggestion was not put forward. The result was a meeting between the Minister of Agriculture, Sir Reginald Dorman Smith, the Permanent Secretary, Lord Perry and Hennessey.

Dorman Smith much favoured the proposal and put forward a Bill to underwrite the extra production of Fordson tractors. There was some opposition to such a Bill from other companies who considered that Fords were being given a monopoly, but since the Fordson was already filling about 90 per cent of the British tractor market, such objections made little headway and the Bill was passed. Thus, when war did indeed begin in 1939—perhaps a little to the Ford Motor Company's relief—tractor production at the Dagenham plant was running high and the Fordson dealers already had considerable stocks of tractors. As had been foreseen by the Ford Motor Company, there was an almost immediate demand for these following the ploughing orders sent out by the Ministry of Agriculture; in the event, the Ministry did not have to pay for a single tractor to honour its underwriting arrangement.

This unique operation, with the foresight and firmness of conviction it required, gave rise to considerable pride in both the Ministry of Agriculture and the Ford Motor Company and it certainly contributed to the resistance against Ferguson and the idea of producing his tractor instead of the Fordson.

In the final analysis, however, practical considerations were bound to win the day and a production switch could only have been made with great difficulty. Even though Harry Ferguson had won the Bank of England over to his side and they were prepared to finance the switch, the government could not sanction it. Even in peacetime it would have been a major operation, and in wartime the tooling and plant resources simply could not be spared for tractor production. Therefore, by the time the May 1940 demonstration of the Ford/Ferguson tractor was given at St Stephens, Bedfont, neither the British government nor the Ford Motor Company were prepared to be moved by what they saw. In fact, to have such a solid obstruction in their path as the government's resistance to a production switch seems to have been convenient for the Ford board of directors. There is no evidence that they ever tried to remove the obstruction, and perhaps it might not have been immovable had Ford and Ferguson wielded the battering ram together. But the Ford directors could see no reason for doing so: their Fordson tractor, though relatively crude, seemed able to meet the needs of British agriculture quite adequately; to have made a changeover would have resulted in a loss of production over several months; to have fought the government in order to obtain the necessary authorisation would have seemed churlish after the government had given them a guaranty that had enabled them to expand the Fordson production; and finally, but by no means least, why should they bother to assist that gadfly that was constantly buzzing in their complacent ears? He might have a good tractor, but his pushing and manœuvring, and his assertive haranguing were quite insufferable.

Shortly after the May 1940 demonstration, Harry Ferguson met with the Ford directors in their luxurious offices in Regent Street. This final confrontation was not what Ferguson would have described as a 'cheery occasion': it turned into a caustic dispute during which some deep-felt accusations of a personal nature were flung out. Ferguson subsequently maintained that, unbeknown to him at the time, the conversation was being recorded. He claimed that some of the remarks passed in the violent atmosphere were afterwards quoted back at him verbatim. This, he considered, was a gross breach of etiquette, for to him, as to most people whose tongues are easily unleashed in moments of anger, it was distasteful to be reminded in a later and calmer moment of the words actually uttered.

Ferguson could not, and indeed did not try to, hide his contempt for the policy of the Ford board of directors. It was summed up in an explosive letter to Lord Perry which began: 'Dear Perry, Some people make opportunities out of difficulties, but you make difficulties out of opportunities.' As he wrote those words, running through his mind were probably recollections of the years of toil and frequent defeats that he had suffered before perfecting the Ferguson System. These people with whom he was dealing were in his eyes too witless to recognise what they were being offered, or if they did recognise it, they were too spineless to overcome a few minor obstacles in the path of accepting it. A Ford executive once commented, 'If you disagreed with Ferguson he thought firstly that you were a fool, and secondly that you were malicious.' This is an overstatement, but it was true that once a person's thinking began to diverge from his own, Ferguson's intractability usually widened the gulf to a point where it was unbridgeable. He was totally unable to put even one toe in the Ford camp to see how matters looked from there, and so he returned to the United States utterly disgusted at what he was convinced was sheer stupidity on the part of the British Ford directors.

17 *Kyes's Trainloads*

In Detroit, in the summer of 1940, Ferguson affairs were not going well either. The reception of the Ford/Ferguson tractor had been ecstatic enough, but the organisational matter of setting up a sales network and arranging for the supply of implements presented many problems. The Sherman brothers, Eber and George, struggled with these and succeeded in solving many of them; but the great euphoria of the months each side of the launching of the tractor in the United States had faded progressively as unsold tractors began to accumulate outside the Rouge River plant. A mighty production line had been established; a fanfare of trumpets had announced the advent on the scene of a revolutionary new tractor; but incredibly this great advance in farm machinery did not seem to inspire much demand from farmers. The reasons were fairly simple: in the early days of the war and before the entry into it of the United States, there was little pressure to increase food production in North America, and nor was there any shortage of manpower on the land to stimulate farm mechanisation. In addition, the Ferguson System tractor faced the same market difficulties as it had in Britain when it was being built by David Brown. Farmers who were already mechanised could not change their tractor for one incorporating the Ferguson System because the implements they already had could not be used with it; they had to change their complete range of implements as well if they wanted to buy a Ferguson tractor. The obvious market to crack open therefore was that of the small farmer who had hitherto remained almost totally unmechanised on his holding. However, the circumstances that were to open the door into that market were not then actual, though they were hovering on the horizon.

The Sherman brothers, despite their good work in establishing a distributor and dealer network, were plainly out of their depth in dealing with the build up of unsold inventory. To add to their difficulties, Henry Ford was scathing about their inability to market the tractor as fast as he could build it. To make his point perfectly clear, he even arranged for a tractor to be parked one morning so that it blocked the entrance to George Sherman's office. The Sherman brothers began to fear that if the situation could not be remedied, the Ford Motor Company might feel compelled to assume the distribution of the tractor, and possibly the manufacture of implements as well. One means of forestalling such a move would be for the Ferguson company to step up its development work on the implement range, and its production by outside contractors, so that in the event of a confrontation with Ford, its possession could be used as a bargaining point. Quite apart from this there was clearly a need to have more types of implement available if the Ferguson System was to gain market acceptance. For example, there was no cultivator in the original British range that was suitable for the inter-row cultivation of maize, one of the most important North American crops. The Shermans therefore instilled a sense of urgency in the small Ferguson engineering team that was working on designs of new implements.

By this time, the engineering team, which numbered about thirty people, was under the immediate control of John Chambers. Willie Sands, unhappy away from his beloved Ulster, had returned there after only a few months at Dearborn. It is difficult to establish whether Ferguson allowed him to go without remonstrating with him at the time, but the fact that he did go was subsequently a contentious point between them. It was the Shermans who decided that John Chambers should be officially nominated chief engineer. He was chosen in part because of his close knowledge of Ferguson's ways and quirks, for they believed that this knowledge would facilitate matters for the rest of the team, and particularly for newly-joined engineers like Ernest Bunting who were unacquainted with Ferguson's exigencies.

In passing, it is worth noting that the fears harboured by the Shermans over the possibility that Fords might want to take over implement design and manufacture seem to have been warranted, because in this period Henry Ford asked George Sherman whether there were not some implements that his staff could develop. Sherman cannily replied that there certainly were and suggested he begin with the combine harvester. 'I knew that would keep them busy for years,' he smilingly told one of the Ferguson engineers. And Henry Ford did employ a man by the name of Kurt Baldwin to experiment with combine harvester designs for a time. (Several years later, Baldwin became well known in this sphere when Allis-Chalmers manufactured a combine harvester that he had designed and which bore his name.)

The Shermans also realised that only the services of an outstanding sales executive could solve the marketing problem that was causing the stockpile of tractors to increase daily outside the Rouge plant. The person on whom they set their sights was Roger Kyes, a young man who was then vice-president of a concern called the Empire Plow Company. Kyes had spent long periods attempting to interest the Ferguson-Sherman Corporation in purchasing sweeps and tines for their cultivators and he was therefore well known to the Ferguson team.

Kyes was a most unusual man; at the time he first made contact with the Ferguson-Sherman Corporation he was about thirty-five; he was 6 feet 4 inches tall and his appearance so ugly—long faced, big nosed and saturnine—that it was immediately arresting. He was outstandingly articulate with measured words and a softly pitched voice, and he was a brilliant salesman. Even Ferguson, before he left for England in October 1939, had been impressed by him. Kyes had given an authoritative talk to the Ferguson engineering team on the subject of the vibrations set up in a cultivator sweep as it was pulled through the soil. He propounded the theory of the usefulness of these vibrations at some length and called their effect on the soil 'chativation', a word that appealed greatly to Ferguson and his engineers. Almost anyone who has worked with Kyes ascribes to him some epithet such as 'ball of fire'. He was indeed ambitious and hardworking in the extreme, but he also had a reputation for ruthlessness that later evoked descriptions such as 'Jolly Roger, the Hatchet Man'. His appearance was probably more to blame for this reputation than were his deeds; for even the wide and brilliant smile that could transform his face, and the braying laugh, were slightly sinister in the suddenness and speed of their coming and going. Kyes's energy and persuasiveness were just what the Shermans needed, so as a start, they arranged to buy some of his time from the Empire Plow Company to assist with their marketing problems.

Kyes's solution to the accumulated tractor inventory was what he called the 'Trainload Campaign'. By some very clever talking, he and Eber Sherman persuaded each of the twenty-nine distributors that had been appointed to take a complete trainload of tractors for sale through their 3000 retail outlets. It was an astute move because the twenty-nine trainloads, apart from clearing the stock, provided an opportunity for good press coverage and public relations, thus laying a foundation for better future sales.

Harry Ferguson arrived back at Dearborn from his abortive stay in Britain, to find the affairs of his company less prosperous than he might justifiably have expected. He was amazed, and angry too, to find that the Shermans had given John Chambers the title of 'Chief Engineer' and he withdrew it at once, stating that in so far as there was to be a chief engineer that person would be Willie Sands. Ferguson never wanted to delegate any engineering authority to anyone, always reserving for himself the final decision over even the most minute detail. And once re-established at the Dearborn Inn, no day passed without a visit to the engineering team in the drawing offices or to the experimental farm close to the Dearborn Inn where field tests were always in progress. He pursued the engineers constantly and often expected them to work extra hours, or at week-ends; but he radiated such infectious enthusiasm for the task in hand that no one objected. Once, on the eve of the Independence Day holiday of July 4th, he went to the drawing board at which Ernest Bunting was wrestling with an implement design. He watched in silence for a few moments. 'Tomorrow's a holiday, isn't it?' he asked, when he had realised that the design could not be finished that day.

'Yes, it is, Mr Ferguson,' said Bunting.

'And I suppose you want to take it, do you?' Ferguson asked pointedly.

'Well, we declared ourselves independent of Britain once and I guess we can do it again,' Bunting replied in his easy drawl. Ferguson appreciated quick wit more than any other form of humour and he laughed.

Most of the engineers enjoyed working with him. They were fascinated by his uncanny gift for knowing what was wrong and by his unerring judgement on a design. (This judgement was intuitive: 'If it looks right, it is right' was one of his favourite expressions.) They may have found his pernicketiness exasperating but at the same time they admired him for the extremely high engineering standards he set. An engineer who presented an implement for Ferguson's inspection and on which the threaded portion on a single bolt protruded more or less than the specified two threads beyond the nut, or on which the end of a bar was not neatly radiused, could expect to have his ears scorched off his head by Ferguson's fulminations. Yet, it was possible, despite his tendency towards intractability, for an engineer to disagree with an opinion that Ferguson expressed; and provided the engineer could argue his point convincingly enough Ferguson would listen. Heated discussions around a drawing-board would sometimes continue at length, for many of the American engineers were uninhibited by the master-employee relationship that characterised Ferguson's dealings with most of his Irish staff. But Harry Ferguson rose to the new circumstances and could completely disarm recalcitrant engineers: he would terminate a discussion that was proving inconclusive by putting his hand on the man's arm and saying, 'Please do it my way—just for me!' The charm of the appeal was irresistible and the engineer would find himself smiling and saying, 'Why, surely, Mr Ferguson.'

The commercial aspects of the business *per se* did not really interest him and the day-to-day dealing with commercial matters was of no attraction whatever. For this reason he was quick to grasp at Roger Kyes as an outstanding asset that the company should acquire and he offered him full-time employment. Kyes succeeded brilliantly in what was a very difficult task. For apart from the marketing problems, there was the implement manufacture to arrange. In retrospect, it is surprising that the Ferguson-Sherman Corporation ever became a viable concern at all when one considers what it set out to do: it was to market a tractor made by another company, and it was to arrange for the production of an implement range to be sold with the tractor. It had very little working capital to carry out this operation, and therefore all the possible implement suppliers had to be persuaded to produce machinery to the Ferguson design without any guarantee of sales, and without any contribution to the cost of setting up the production facilities. This objective was bold, even brazen; the fact that it was achieved is eloquent tribute to the qualities and enthusiasm of a small group of men. So persuasive were Ferguson, Kyes, and Eber Sherman, that manufacturing companies one after the other fell under the spell of the vision that they presented. This vision was one of hundreds of thousands—even millions—of implements that would be necessary for use with the hoards of little grey tractors that would soon be effortlessly tilling the farms of the world, sweeping away laboriously plodding horses, mules and other draft animals. It was an attractive ideal, and Ferguson and his close associates were so deeply inspired by it that they were able to carry many people with them—people who in sufficient cases were manufacturers who agreed to produce Ferguson implements.

The arrival of Kyes as a full-time employee of the Ferguson-Sherman Corporation coincided with an incipient souring in the relations between Ferguson and the Sherman brothers. At the root of the trouble was that George Sherman took to airing his views on engineering to an extent that irritated Ferguson. It is not on record how good or bad his views were, but since Ferguson was adept at recognising and snatching up good ideas when he saw them, one must conclude that George Sherman's on tractor and implement design were mediocre. His fault was that he would not recognise the fact and continually intervened in engineering discussions. As one of the engineers said at the time, 'If only George would learn to keep his mouth shut he'd be all right'. But he did not learn and the situation deteriorated to the point where he was forced out of the Corporation. Ferguson maintained later that he had never trusted George Sherman's judgement or ability since the Evansville plough days of the 1920s, and that he only agreed to have him as a vice-president of the Ferguson-Sherman Corporation 'to give him a second chance and to please Eber'.

After George Sherman's demise, it was almost a foregone conclusion that Eber would follow the same path sooner or later. Kyes was thrusting vigorously ahead, crunching through problems with the resolution of an ice-breaker. At the same time, he was becoming ever more appreciated by Harry Ferguson—and also by Henry Ford who had known him since childhood through family friendships. Kyes caused much disquiet among older Ferguson employees however; for, unlike an ice-breaker, Kyes did not always confront his problems head on but often manœuvred intricately around them before battering them. In fact his cleverness seemed to many to have a Machiavellian streak in it, and for this reason it inspired uneasiness. One engineer was so disturbed by the trend of events, as he saw them, that he asked John Chambers to speak to Ferguson about it.

'Mr Ferguson says don't worry,' was the message that Chambers brought back; then he added, 'and if it comes to conniving, Mr Ferguson is a match for anybody!'

With each passing week, Kyes obtained a firmer grip on the affairs of the Corporation and his dynamism overshadowed Eber Sherman. It was characteristic of Ferguson that once he no longer had any use for someone he created an atmosphere in which it was all but impossible for any self-respecting person to stay on. The process was gentle, but ruthless at the same time, and in 1941 Eber Sherman resigned from the presidency of the Ferguson-Sherman Manufacturing Corporation. The name of the company was then changed to Harry Ferguson Incorporated and the hard-driving Kyes had a firm grip on the controls of management.

As we have already seen, a well-refined sense for sales promotion was brought to bear in the planning of any Ferguson operation. Many of the tenets established in the Ferguson-Brown days were transferred to the North American scene. For example, great emphasis was laid on product training, and part of a dealer or distributor's franchise agreement was that at least one or two members of the staff had to go to the Ferguson school at Dearborn before the company was allowed to handle a single Ford/Ferguson tractor. Ferguson also made numerous speeches that were calculated to attract press attention. In December 1940 he told a large gathering that three hundred thousand tractors should be shipped to Britain in order to enable 1,051,000 horses to be taken off the land, thus freeing an extra five million acres for human food production. According to his calculations, this would free two thousand ships that would otherwise be carrying food to Britain, and only forty ships would be needed for the tractor shipments. The equation of farm machinery with social progress was one that Ferguson was now beginning to propound with increasing emphasis, and it was one that many journalists noted and commented on with interest.

Kyes and Ferguson together formulated and promoted one scheme that was very subtle for the early 1940s, even if by today's standards of highly sophisticated public relations it would be undistinguished: they created a National Farm Youth Foundation. The Foundation awarded scholarships, twenty-nine of them initially, under which selected young people were given a year's course in farm management and engineering. Most of the course was at the agricultural college nearest to the award winner's home, but one month of it was at Dearborn where the Ferguson staff gave the students a complete course in the Ferguson System. There was no overt indoctrination—Ferguson was too subtle for that—but the net effect was that even without indoctrination the young person was turned into a disciple who would go back to his home area. In addition, the nominations for the scholarships were made through the Ferguson dealers and distributors, and inevitably this raised the status of these companies in the eyes of the local farmers whose sons were in the running for such grants.

The award of the scholarships could be exploited for publicity purposes, and occasionally Ferguson made a point of doing so himself. In August 1941 he went to Portland, Oregon, to present plaques to the dealers who had sponsored two winning candidates for the Foundation's scholarships. During the course of a speech he made to the three hundred and fifty people gathered for the occasion, he said: 'The cost of living must be kept in check. We are really in serious danger of a disastrous inflation that might mean that all we have will go up in smoke. It's clear that if the cost of living continues to increase, the cost of the defence programme will be colossal. The answer lies in farm machinery. It must be used to reduce the cost of food. There are six million eight hundred thousand farms in the United States, but only about 800,000 of them are mechanised.' This was one of the first occasions on which Harry Ferguson expressed in public the kernel of an idea that was beginning to obsess him. He later developed it still further and preached it as a doctrine.

Under the firm and compelling hand of Roger Kyes, the business of Harry Ferguson Incorporated began to gather momentum. Kyes's efforts were, however, fortuitously assisted by a change in the circumstances of farmers in North America. Even before the

Japanese attack on Pearl Harbor, the effects of the war in Europe and North Africa had begun to make themselves felt on the United States economy. But with the American entry into the war there was a demand for higher agricultural production with reduced manpower and this finally brought the conditions favourable to the introduction, on a relatively large scale, of the Ferguson System. The need for farm mechanisation coupled with the factor of Kyes's dynamic efforts, pushed the tractor sales in 1941 to about 40,000 units; and total turnover of the Ferguson company, including sales of implements, accessories and parts were just short of $27 million.

18 *The Loss of a Friend*

At last, after so many years of disappointments, Harry Ferguson was seeing his dream fulfilled; tractors incorporating his system were in large-scale production and he had a company with a big and growing turnover. Unfortunately, the blossoming of this success was marred for him by the death of John Williams. Williams had rejoined the Royal Air Force at the outbreak of war and one night, driving his car with headlights almost entirely masked to comply with blackout regulations, he drove into a stationary and unlit truck. Ferguson was stricken by the news and decided that some tribute or memorial was necessary. It so happened that during September 1940 he had been 'thrilled and filled with admiration for our (R.A.F.) successes' during the Battle of Britain and had wanted to contribute to the cost of a Spitfire. The *Belfast Telegraph* had a fund for just this purpose and on hearing of Williams's death, he sent a telegram to the newspaper. 'I subscribe £2000 in grateful appreciation of the R.A.F. and in memory of my greatest friend, a very gallant gentleman, Flight Lieutenant John Lloyd Williams.'

The death of a close friend must inevitably have short- or even medium-term repercussions on the life and attitudes of the survivors, but in the case of Ferguson the repercussions echoed for the rest of his life. He probably never realised it, but John Williams was more than just a very good friend: he was a stabilising influence in Ferguson's life. Ferguson's periods of energetic activity during which no obstacle whatever could deflect him from his drive towards pre-set objectives were interspersed with short periods of bleak despair. This, in fact, is often a characteristic of men who aspire to, and reach, great heights of achievement. Today, people who suffer from it are termed manic-depressives by psychiatrists, but a brief examination of the lives of some of the outstanding geniuses of history show that they too suffered from the condition.

One can conclude that many manic-depressives are so much

favoured during their periods of imaginative elation, energy and enthusiasm that they become supermen capable of accomplishment far superior to that of more ordinary people. The periods of depression they suffer are the penalty they pay for their creativity at other times. Some medical authorities believe that there is a pattern in the lives of most manic-depressives. In their early working years their buoyant periods are almost continuous with only brief spells of dejection. With increasing age the periods of depression encroach increasingly on the generally animated character of the person. Gloom strikes at them more often and for longer periods, and the process can be so inexorable that in old age they are in an almost continuous state of depression illuminated by only brief flashes of their old energy and creative thinking. Another characteristic of manic-depressives is that they often suffer from psychosomatic ailments; indeed, such ailments may be the factor that first leads a doctor to suspect that he has an incipient depressive on his hands.

Ferguson's relationship with John Williams, dating from the early years of the century when they met at the Belfast Technical College, was a very special one. 'Cap', as Williams was affectionately called by all the Ferguson employees in Belfast (a nickname derived from his World War I rank of captain), was a sheet anchor for Ferguson. He was unfailingly cheerful, a great joker and teaser, and he had the key to Ferguson's limited store of humour. In times of great adversity and when Ferguson was working himself into a state of stress, Cap could produce one of his witticisms or tease Ferguson into taking the situation and himself less seriously. It is significant that there was much similarity between Maureen Ferguson's veneration for her husband and the attitude of John Williams towards him: Williams's sister was always surprised by the way in which Cap, despite his far superior education and background, willingly subordinated himself to Ferguson. Had this not been the case, it is doubtful whether their intimate friendship of nearly forty years could have lasted. Together, Maureen Ferguson and Williams were a powerful force in keeping their loved but difficult prodigy on an even keel. Their combined efforts were able to help him through his more difficult periods, but with the loss of Williams there was no one left to tease and amuse Harry Ferguson towards a more balanced view of matters. Maureen Ferguson alone, despite her adulation and her consideration for her husband's every need, could not provide the light relief that Williams had, or really appreciate their kind of humour. During the witty exchanges that Ferguson and Williams often had, she remained unmoved by their laughter.

It must not be overlooked that by the early 1940s Harry Ferguson was approaching his sixtieth year. He had suffered from chronic insomnia for much of his adult life, as well as from stomach trouble that was certainly, seen in retrospect, of nervous origin. Following John William's death, there was a marked increase in his ailments and disorders; his eyes in particular pained him and he took to having all letters and documents read to him. His doctors ordered total rest on several occasions during the war years in the United States. Ferguson was incapable of following such orders for he had no hobbies and so he merely fretted while away from his work. While he was suffering one three-month holiday in Maine, upon order from the Mayo Clinic, he summoned John Chambers from Dearborn every other week-end in order to be brought up to date with the progress being made by the engineering department in its development of new implements. While Chambers was with him he was animated and happy, but for the rest of the time his enforced idleness made him even more miserable than the stress of work. He therefore escaped back to Dearborn as soon as he could.

There were quite frequent dealer and distributor meetings at Dearborn, and the main purpose of these gatherings was to create a solid team spirit, and an ardent enthusiasm for the Ferguson System. Harry Ferguson excelled in creating an almost mystical aura around it and what it could do for mankind. His charisma was spell-binding; realists and cynics found themselves convinced, against their better judgement, of the immense value to the world of this little grey tractor. To read the texts now of speeches that Ferguson made during those years in America leaves one with a sense of bewilderment; one cannot at first understand how such bold claims were not met with eyes rolled heavenwards in disbelief, or with smirks betraying the inner thought that here was the charlatan of all time. But no, those speeches made abiding impressions, reinforced too by the excellent demonstrations that usually followed them. Harry Ferguson was no great orator, and his power to convince certainly did not come from the well-turned phrase and the perfect timing of a Winston Churchill, or from the imposing manner and sweeping gestures of a Charles de Gaulle. They came from nothing other than the depth of his own conviction and earnestness. These allowed him to do what were basically ludicrous stunts in public with such aplomb that no one thought them ludicrous at the time. For example, he once appeared before a large audience of his dealers and distributors with black grease smeared over much of his face and his clothes awry. He began his speech, but not surprisingly many of the audience were exchanging asides about the unusual appearance of the normally immaculate man they knew.

'You see, gentlemen,' he said, suddenly breaking off in his speech, 'you're not paying attention. You are distracted because I am dirty and untidy. And your customers will be distracted too if you and your premises are dirty and untidy. Let this be a moral lesson to you.' Having said this he whisked out of the room to clean himself up and came back to begin his speech again.

One of the great question marks that must forever hang over this period and the unique arrangement between Harry Ferguson and Henry Ford is to what extent one man influenced the other. Whose were the original ideas that crystallised into the doctrine of the benefits that the Ford/Ferguson tractor could bring to mankind? Did they spring from one or other of the two men alone, or were they the result of a cross-pollination of two idealistic minds? Unfortunately no one was privy to the many intimate conversations between the two men. The only thing that is certain is that both men saw the new machinery as a steel Messiah that would save the world from many of its evils. Ernest Bunting and Ed Malvitz one day gave a demonstration to Henry Ford of some weedhooks that they had developed for the Ferguson plough. Bunting and Malvitz were very worried that the demonstration would fail completely; it had just rained and the clayey soil near the Rouge River stuck to the mouldboards in the trial they made just before Ford arrived. But Malvitz took a flying run and the tractor and plough completed one length of the furrow in good order with the mouldboards scouring well and the shoulder-high vegetation being perfectly buried with the help of the weedhooks Bunting had designed. They expected Ford to pass some comment on the excellence of the work, but instead he walked slowly to the front of the tractor and laid his hand on the bonnet. 'You know, boys,' he said, 'this little tractor is gonna put an end to all the wars in this world.'

The question of whom led who into such idealistic conceptions is indeed an intriguing one. Undoubtedly, Ferguson introduced the idea that his inventions could bring great benefit to mankind during the meetings and demonstrations that he and Ford had together in the autumn of 1938. Indeed it was this concept that initially helped him to win Ford's collaboration in the joint venture. Ferguson was wily enough to realise also that how and when he presented the idea to Ford would be instrumental in deciding whether the older man would be impressed by it. The persuasiveness of his argument was therefore certainly enhanced by the intimate atmosphere generated between the two men as, following the demonstration of the tractor, they strolled side by side in the crisp, golden afternoon. The good earth beneath their feet, the piercing clarity of the sky above them, the hues of autumn in the trees, and the glistening furrows recently turned by the little tractor, now parked silently on the headland—all these set the perfect scene for a serious talk about the richness of the land and how its wealth could be unlocked for the betterment of mankind through the use of farm machinery. A little later, as they sat alone at the small table brought to the field for them, the accord between them matured like ripening fruit. Ferguson was well aware of the role the surroundings were playing in reinforcing his arguments; years later, he wrote in a letter: 'The choice of a time and place for our conversation was as important as what was said.' And we can be quite sure that he was responsible for that choice.

Later, however, other ideas were incorporated in the philosophy of what the Ferguson System could do to improve the world; and it seems likely that a good deal of Henry Ford's ideas rubbed off onto Ferguson during the course of their private talks. The concept that inflation is one of the major menaces in the world, and the idea of checking or even reversing it by providing cheaper food through farm mechanisation, could well have been a compounding of the two men's ideas. Certainly, the corollary that the machinery itself would have to be cheap in order to play a part in lowering food prices, and to achieve the sales volume necessary to make a world impact, was one that contained much of the Ford doctrine. For did not Henry Ford once say: 'Every time I reduce the price of my car by a dollar, I get a thousand new buyers.' And did he not also say: 'The man who can produce a car which is entirely sufficient mechanically and whose price is within the reach of millions who cannot afford one now, will not only grow rich but be considered a public benefactor.'

If Henry Ford believed this about cars, its application to farm machinery was even more rational because of the more basic role of such machinery in the fabric of economic life. Ford embraced the idea enthusiastically as is witnessed by his material support of Ferguson; but it could well be that it was much of his own thinking that was incorporated into the concepts that Ferguson voiced. That the two men had a genuine understanding and respect for each other, and that they stood on common ground in their ideologies, is beyond doubt. They had their fights, and they were usually the senseless ones of stubborn men unable to see the other's point of view. One source of friction occurred over the trifling matter of the production of a British aero engine in the Ford plants: before the United States entered the war, the British wanted the Rolls-Royce Merlin, the engine that powered most of Britain's and some of America's war planes, to be produced in North America. Ford was asked to take it on but refused. Ferguson was incensed at what he considered a betrayal of his country, and even though it had absolutely nothing to do with him, he told Ford that it was his duty to produce the engine. Ford in his turn was irritated by Ferguson's continual representations for a place on the management board of the British Ford Motor Company. But despite these disagreements, the personal relations were normally good. Ford even went to the trouble of having a sack of flour specially ground at the old mill in his Greenfield Village Museum when Ferguson complained to him about not being able to get wholemeal bread for his breakfast. And Henry Ford tried to involve the Fergusons in his square dancing evenings at Greenfield Village, for Ford himself was a folksy supporter of traditional American activities. Ferguson was totally bored by such occasions, though his daughter Betty quite enjoyed them and has vivid memories of being twirled about by the aged wraithlike figure of Henry Ford. Ferguson's main enjoyment of Ford's company was certainly the philosophising in which they indulged. 'I spent several wonderful years with Mr Ford discussing the improvement of the world by eliminating draught-animals on the farm,' he later declared. Some sources, notably Kyes, maintain that personal relations between the two men was full of friction and that their agreement was on the point of foundering many times, but it is doubtful whether they really ever deteriorated to that degree.

19 Steel Crisis

Ferguson made much of his own good fortune by his brains, hard work and perseverance, and he also made much of his misfortune by being too intolerant and cocksure, with the result that he aroused animosity. However, the near disaster of steel and other material restrictions that befell his embryonic company in 1942 was no fault of his. The U.S. government was forced to impose quotas of material on all manufacturing companies and direct the scarcer commodities to the quarter where they were most urgently needed for armaments. The allocations to companies whose products were not considered to be of top priority were based on a percentage of the quantities of materials used by each company in the years just prior to the war. Whereas for most companies this did not cause excessive hardship, for Harry Ferguson Incorporated it could have meant total catastrophe. The company had no real prewar production base on which it could claim allocations of steel and other materials, as it had only begun operations in 1939 and reached reasonable production levels in 1940–41. The marketing problems had barely been solved, therefore, when material allocations for tractor production dwindled and finally dried up almost completely. In the latter half of 1942, the tractor production line at the Rouge plant was completely idle for almost six months. Whereas by December 1st, 1941, 90,000 tractors had rolled off the line, the next 10,000 were not produced before late October 1942. From a total turnover of nearly $27 million in 1941, the figure dropped to $14 million for 1942 as the marketing problems gave way to ones of production.

The differing tactics adopted by Kyes and Ferguson to overcome the shortage of materials are interesting. The practical Kyes, accustomed to thinking at the detailed level, immediately launched the idea of having all Ferguson distributors and dealers look for old Fordson tractors that could be broken up and re-smelted for the production of new Ford/Fergusons. Many

Fordsons were removed from the nettle patches around farm buildings and transported to Detroit for scrap. Kyes also used his many contacts in industry to wheedle second-grade ingots out of the steel mills. And because copper was in even shorter supply than steel, he persuaded the Ferguson and Ford engineers to produce an austerity version of the tractor with no electric starter motor and with a few other simplifications.

Ferguson, on the other hand, believed that some ardent lobbying would be the best way of obtaining greater generosity from the U.S. government. As usual he was thinking of overall strategy and concepts while others worked out the details. He managed to exert sufficient pressure to have a Senate Investigating Committee see a demonstration of his equipment. The Committee, which included vice president Henry Wallace and Harry Truman, was impressed by the machinery and by the earnest presentation made of it by Ferguson. From there, it was not too difficult to strike out for the top and persuade President Roosevelt that he too should see the tractor and implements at work. Ferguson was convinced that if he could in person explain his System and what it could do for mankind to the President, the difficulty over material allocations would be solved. Roosevelt agreed to let Ferguson give a demonstration on his own farm, Hyde Park, and came to see the equipment at work. Ferguson put on his smoothest display and used his persuasiveness to such effect that Roosevelt bought a tractor and range of implements and, more important, promised to ensure the necessary supplies of materials for the resumption of large-scale manufacture. It was a fine hour for Harry Ferguson, one which he had created himself because neither the Ford Motor Company, nor Henry Ford, nor Edsel his son, liked dealing with politicians; Ferguson, who had many of the attributes of a clever politician himself, did. Quite apart from his belief that it was through influencing top people that one got one's desires fulfilled, he also liked consorting with them; though he was not a social climber in the conventional sense, nor over prone to name-dropping, he obviously derived pride and pleasure from such acquaintanceships and sought to create them.

Some of the distributors of Ford/Ferguson equipment became personal friends of the Ferguson family, and developed quite close personal ties with Ferguson himself. None were ever able to replace John Williams in his affections or reach such a degree of intimate friendship, but nevertheless they got on well. The distributor who probably became best known to Ferguson was Fritz von Schlegel of Pasadena, California. Von Schlegel had been a distributor for the British-built Fordson tractor that had been imported into the United States by the Shermans through their Sherman and Shepherd Company. The Fordson was not well suited to the row crop farming of California and von Schlegel did little business with it, despite his well-run organisation. When, therefore, the rumours of a new Ford-produced tractor began to seep out of Detroit in 1939, von Schlegel was among the first to cock his ear with interest and travel to Dearborn to see the prototype Ford/Ferguson. At that first demonstration, von Schlegel, who was always outspoken, voiced the opinion that the rear-mounted cultivator would be no use for row crop cultivation. The whole team of Ferguson engineers, on hearing the remark, glanced nervously at Ferguson and prepared to run for cover in case the fuse of his rage had been lit. Luckily he had not heard, but the engineers' obvious fright conveyed very well to von Schlegel that such comments were best left unsaid.

Fritz von Schlegel later heard that the Fergusons would like to take a holiday in California, and was asked whether he could arrange it. Von Schlegel made reservations at the Del Monte Lodge on the Monterey Peninsula. There the Pacific surf strikes a coastline of outstanding grandeur, the spray soaring as it shatters on the rock outcrops and foams into the sandy coves; pine and cypress groves and green fields run down to the sea; on many of the promontories contorted tree trunks whitened by sun and salt-laden wind lie or lean weirdly among the rocks, smoothed and shaped by the forces of nature.

In the luxury of the Lodge, in this beautiful area, Harry Ferguson fretted for over a month of so-called holiday. Fritz von Schlegel drove up from Pasadena regularly to be with the Ferguson family and to try to make their stay a happy one. He believed that Ferguson would be interested in seeing some Californian agriculture, especially that near Salinas (in Steinbeck country).

'We'll drive over to Salinas and I'll show you how they farm,' von Schlegel once said.

'No, Fritz, they're not doing it right. I don't want to see it. Anyway, I'm feeling tired,' was the reply. Ferguson always had an excuse for not wanting to go onto Californian farms with von Schlegel, though von Schlegel was certain that he later went alone to have a careful look. It may be that Ferguson was just depressed and genuinely uninterested when von Schlegel suggested the visits, but a likelier reason is that Ferguson knew that his rear-mounted cultivators were not ideal for the row crop work around Salinas; yet mid-mounted cultivators of the type that would have been completely satisfactory for the specialised agriculture of California did not fit into the Ferguson System. To have a direct confrontation with this fact would have challenged his own belief in the complete and universal suitability of the Ferguson System for all types of agriculture. Its limitations, even though fairly insignificant, were not something that he was prepared to accept in the company of von Schlegel, or anyone else, even if he admitted them to himself.

After a number of relatively gay occasions, one of them a ranch barbecue in the Carmel Valley—which appealed to Ferguson's interest in 'things Western'—he took von Schlegel aside. 'Fritz,' he said, 'I should like to help you and do you a big favour now that we have become such good friends. I'll arrange a big demonstration in your area. Next year in Arizona.' What he had in mind was to move his engineering team south from Detroit, as in the previous winter, and terminate several weeks of field testing with a demonstration for U.S. distributors, and of course for local dealers and farmers.

Fritz von Schlegel did not know what being host to the Ferguson circus in Arizona was going to involve, but in January 1942 the team arrived at Phoenix. Von Schlegel had made arrangements for the Ferguson family to stay at the Camel Back Inn, one of the most famous and beautiful winter resorts in the United States. A few miles out of Phoenix, it consists of a cluster of white buildings set in a green jewel of palms and lawns that sprout abruptly in the unlikely and prickly surroundings of the Arizona desert. The luxurious appointments, the beautiful gardens and swimming pool, the desert climate in winter with its crisp nights and warm sunshine during the day enrapture most people, and it was a perfect choice for someone as difficult to please as Ferguson—not that there is any record of his having been enraptured by it.

Harry Ferguson was feeling run down again and was difficult to deal with. One of his staff assembled a particularly elaborate implement, involving literally hundreds of nuts and bolts, and just as he finished his half day's work, Ferguson came over and gave it a piercing scrutiny. 'You didn't put any grease on the threads,' he said accusingly.

'No, Mr Ferguson, because it's only going to be used for the one demonstration and then dismantled again.'

'Take it apart now and put it all together again with grease on every thread,' Ferguson snapped.

He was so edgy that his men were even more respectful towards him than usual. When they were in the field they had a shallow hole in the ground prepared so that they could throw their cigarettes into it and scuffle them over as soon as someone warned that Ferguson's bright green Lincoln was in sight. Smoking in working hours was always forbidden by Ferguson, but he would normally turn a blind eye to a cigarette butt on the ground.

But despite the vicissitudes and irritations of dealing with Ferguson, the demonstration went brilliantly, except that Ferguson used the term 'American peasants' during his presentation, a description that did not evoke a very favourable response from the mainly prosperous farmers present. After the demonstration, Ferguson went back to California with von Schlegel to see a doctor, but nothing was found wrong with him. The doctor did, however, comment that Ferguson was run down and needed a rest.

20 *World Plan*

The agricultural climate of the United States underwent a major change in the year 1943; indeed there were also at that time the first stirrings towards a world-wide change in agriculture. In the United States itself, the effects of the war-based economy were biting into everyday life, and there was a premium on farm machinery. The lobbying that Ferguson had done in Washington, coupled with Kyes's contacts with steel companies, resolved much of the difficulty over material allocations that had afflicted the Company in 1942. In addition, the Ferguson company did some neat manœuvring in respect of other companies' steel allocations: some of the large companies found that it made poor economic sense to manufacture the stingy few thousand tractors authorised for a period of several months when in all probability their production plants were already working close to capacity on war equipment. Ferguson, Kyes and other top executives of the company played on this and in many cases were able to persuade other companies to make over their farm machinery material allocations to Ford and Ferguson. In this way, the Ferguson company cleverly worked itself into a favourable position in the North American tractor and implement market during the remaining war years. In 1943, total turnover of the Company rose to over $19 million, and for 1944, it reached just under $50 million.

In May and June 1943, there arose an excellent opportunity for Harry Ferguson to propound his philosophy. At Hot Springs, Virginia, at that time, the work of a few dedicated men culminated in an International Food Conference. The movement towards this event had begun in the 1930s when a group of people at the League of Nations in Geneva began to arouse interest in nutrition and in human dietary needs for optimum health. The linking of diet to health was an innovation to many people. It was then a natural step to link nutrition with agriculture, for farmers in the 1930s were constantly finding themselves in difficulty owing to the inability of depression-stricken populations to buy their produce. The League of Nations group included Stanley Bruce, one-time Prime Minister of Australia and then High Commissioner in London, his economic adviser F. L. McDougall, and a Scottish doctor, John Boyd Orr. They believed that the world need for more food and the desire of farmers to produce it as efficiently as possible could be effectively linked and start the world on the road to economic abundance—thus killing once and for all the economy of scarcity that obsessed men's minds.

In the immediate pre-war years, the relation between nutrition, health, and agriculture became more fully recognised. Nutritional standards were established and, oddly enough, the Germans and the British used them in their wartime rationing plans. It is even odder that rationing in Britain produced a better level of health among the people generally than had ever been known before: the rich fared less well than was their wont, but the poor were for the first time eating balanced diets. The ration book provided in effect a simple lesson in nutrition.

The war, however, caused the cancellation of several international meetings on the subject of nutrition and agriculture that the League of Nations group would have held had they not been scattered. One of them, McDougall, went to the United States as Australia's representative in the negotiations for an international wheat agreement. During the negotiations McDougall met with several other delegates outside the main sessions to discuss matters of even greater import and breadth, in particular world food problems. McDougall also wrote an article, published in the *Annals of the American Academy of Political and Social Science*. He stated in his article that the reconstruction of the war-shattered world would hinge largely on food, owing to the privations during the war. He also stated that, since upwards of 60 per cent of the world's population was engaged in farming, the improvement of agriculture and farmers' living standards was an outstanding social problem. He therefore proposed that an international organisation for food and agriculture should be established as a first step towards a world agency for peace and the improvement of man's lot. Mrs Roosevelt happened to get a copy of the article and was intrigued by its rationale. She gave it to her husband to read and he, already preoccupied with plans for preventing a repetition of the catastrophe of war when the then current conflagration had been extinguished, was sufficiently convinced to call a conference at Hot Springs in May 1943 at which delegates from allied countries discussed food and agriculture on a world-wide scale. At the same time they discussed the need for an international organisation to assist the development of agriculture and increase food supplies. There was much basic agreement among the delegates and in effect this meeting was the conception of the Food and Agriculture Organisation that, after a two-year gestation period, was born at Quebec in 1945. With John Boyd Orr as its Director General, F.A.O. was the first of the specialised agencies of the United Nations.

That first conference at Hot Springs in 1943 attracted about a thousand delegates from forty-five countries, an expert and informed audience on agricultural matters, as Harry Ferguson realised. He therefore decided to give a series of demonstrations in the hope that the delegates would appreciate the worth of his inventions and return to their countries with an impression of their excellence; after the war they would, he hoped, retrieve these impressions from the storage of their memories and influence governments to use the Ferguson System in the reconstruction of the world's agriculture.

He prepared the speech he planned to make at those demonstrations with infinite care. He often stated that if an engineer should spend so much time and energy in considering the type and placement of a nut or bolt, someone responsible for speeches and writing should take equal pains over a single word. At those demonstrations, held at Bethesda, Maryland, Ferguson gave the first complete exposé of the dreams that inspired him, of the beliefs that obsessed him.

'Gentlemen, we have not asked you to come here primarily to see a tractor and a new system of implements,' he told the delegates. 'That is a secondary reason for the invitation we sent you. Before I go any further, I want to impress most deeply upon you that it is the *Plan*, the idea behind this new System, which is of primary importance.

'I suppose you know that only one invention in ten thousand succeeds. The reason is that there is no plan behind the failures.

The one that succeeds has a plan behind it. If we had no plan behind our machinery, it would also fail.

'You would not come here and take back to your respective countries any possible story or report of the worth of this new machinery unless I gave you a background upon which to judge it. Therefore, I am going to tell you the Plan on which we work. When you know what that Plan is, I know you are going to approve of it. I know you are going to say it is the most far-reaching Plan that man has devised. If you agree, then your task of judging the worth of the machinery is simplified because you will have a standard by which to judge it. The machinery is of secondary importance, but without it the Plan could not be carried out.

'First, there is the Plan. To put that plan into effect, machinery must be designed fitted to that plan. If the Plan be good and the machinery be good, then we have the greatest and best news that you have ever heard. We have a new hope for mankind.

'Never for a moment, during the twenty-six years I have been working with some of my friends in Ireland and over here, have we devoted our attention to any form of localised machine; that is to say, we took into consideration every world condition that exists. No country has been left unconsidered.

'I made up my mind all those years ago that unless we could make a machine for the whole world, which would revolutionise agricultural production all over the world, we would fail in our task in just the same way as many big companies have failed in the past. All the companies making tractors today are scarcely touching the fringe of the possible world tractor business. In other words, the farm mechanisation industry is hardly born yet. You can judge, today, this plan and this equipment from the standpoint of world conditions.

'The abolition of poverty is our first problem. What is poverty? What is want? Poverty and want can best be described as the inability to purchase the bare necessities of life. Any man who cannot obtain the necessities of life is definitely in poverty. Any man who can purchase them is not in poverty. He may be poor, but he is not in poverty.

'Many plans have been presented for meeting the problem of poverty and destitution. But none of them can possibly succeed unless something is done to create the wealth with which to put those plans into effect.

'I think my own British countryman, Sir William Beveridge, has suggested a very human plan. It has many excellent qualities and it is one which I personally support up to the hilt. It is stalled at the moment, however, and will remain stalled until the British government can find some way of financing that plan. The problem is to find that wealth. Sir William proposes something like two pounds a week to meet the problem of poverty—approximately ten dollars. Very good! But what good will the proposal alone do?

'Will that money keep people out of poverty? What do two pounds or ten dollars per week mean? Two pounds or ten dollars will keep a man out of poverty provided it will purchase for him the necessities of life. But he will still be in poverty if it won't. (Only) one pound or five dollars will keep a man out of poverty provided the necessities of life are cheap enough. You may keep that man out of poverty for just the exact length of time that it takes for the prices of everything to go up, and then he is in poverty again.

'We cannot, and never will, solve the problem of poverty until we solve the problem of stabilising prices and until we solve the problem of creating wealth to a far greater extent than we have ever done in this world before.

'When this war is finished, there will be two ways in which we can run the world. We must either progress or go backwards. We cannot stand still. One way to run it is the way we have been running it. And the way we have been running it is on the basis of every-increasing wages, salaries, and profits to meet the ever-increasing cost of living. In every country we continually are raising wages; shortly following these raises, up goes the cost of living again on all the things that we buy, and then wages must go up again.

'That can only lead to disaster and further wars. There can be no peace in this world, no security, no real solution of the problems of want and poverty, until we stop this ever-increasing cost of living and ever-increasing rise in wages and salaries. You do not make a man any richer by giving him an increase in wages, except as a purely temporary expedient. In a little while that increase in

Harry Ferguson's parents

Harry Ferguson in 1908

Harry Ferguson with his aeroplane at Magilligan in 1910

An early Ferguson monoplane, with the designer at the controls

Flying at Magilligan, 1910

The Ferguson family with Harry's plane, 1911–12

Harry Ferguson and John Williams, 1911
Maureen Ferguson

Harry Ferguson and 'Leslie' in racing Vauxhall, 1912

May Street garage; Vauxhall races with 'Leslie', 1912

Before the T.T. races; Harry Ferguson with Lord Curzon and Captain Campbell, 1928

Ferguson tractor at Chartwell; watched by Winston Churchill, Anthony Eden and Christopher Soames

John Chambers, Archie Greer, Willie Sands and Harry Ferguson with a 1937 tractor

Charlie Sorensen, Eber Sherman, Harry Ferguson and Henry Ford at Dearborn, 1939

Harry Ferguson explaining a model of his tractor to Henry Ford in the field at Dearborn

Harry Ferguson at Abbotswood, 1948

Harry Ferguson drives his tractor down the staircase at Claridges, 1948
Harry Ferguson in the vegetable garden at Abbotswood, 1949

Harry Ferguson driving his tractors at Abbotswood, 1950

Harry Ferguson leaving Claridges after the Ford lawsuit meeting, 1949

Miniature of Maureen Ferguson by John Henderson

Harry Ferguson in the library at Abbotswood, 1955

wages has put up the cost of living and he is just as badly off as he was.

'Take, for example, the very serious coal strike which we have had in this country. The miners wanted two dollars a day more than they were getting because the cost of living had gone up. We will assume that they get their two dollars per day. Who benefits? Half a million miners and their families. Bear in mind that coal is in direct or indirect use everywhere, every day, all the time, and has a direct bearing on the cost of everything we eat, wear and use.

'Half a million miners and their families, some two or three million people, get a temporary benefit by this increase in wages. For how long? For a brief period since it is only a temporary expedient. Two or three million people are benefited for a short time and then they are just as badly off as ever. But there are another 130,000,000 million people who are immediately and definitely worse off.

'In another few months, the miners will be shouting for another rise. War or peace it is always the same. They will have to get it and the cost of living will go up again. That is the system on which we have been running the world for many years, and it is the system which we will have to continue, unless someone does something to make a change.

'Our new "Price Reducing System" is a complete revolution. It alters the whole outlook for mankind. Let's apply it to the coal-mines. Suppose President Roosevelt could say to us today, "Go out and make new machines that will bring down the cost of living by $2 a day", and suppose we did. What would happen? The miners would be perfectly satisfied. They don't care whether they get $2, or whether they can buy the equivalent of $2 more with their present wages. The miners would get what they want with a price reduction the equivalent of $2 a day, and there would be an immediate and welcome benefit to the other 130,000,000 people in the country. Instead of having nearly the whole population worse off when any one section or group receives a rise in wages, you would have the whole population immediately benefiting.

'Gentlemen, this is the way to peace. This is the way to prosperity and the future of the world.

'But there is only one place to start. It is the farm which is the basic cause of the "Price Increasing System". The whole trouble is definitely in agriculture. This is not the farmer's fault, as we shall shortly see. He must be given the opportunity of producing the necessities of life at a cost so low that we can cut the vicious circle of ever-increasing prices and wages.

'We are going to show you how to cut that vicious circle and bring about the benefits of the "Price Reducing System". We have made a complete revolution in the application of mechanics to the land.

'The methods of production on the farms of the world today are hopelessly antiquated. Look at the equipment with which farmers are trying to make a living. You will be shocked to see the equipment with which the farmer is working his farm. You will see why the cost of living is always going up.

'In the past fifty years the world has progressed more than it did in the 50,000 years previously. Why? Because the inventive genius of the world turned its attention to industry for producing automobiles, radios, etc., and devoted its time to making equipment for factories to produce at ever-reducing cost.

'While all this genius has been devoted to the manufacturing industry, it has not been applied to the farm, and the farmers' equipment today is substantially as it was hundreds of years ago. There really has been very little advance.

'There are, I suppose, only two or three million tractors in the whole of the world, but there are still many millions of horses, mules and other draft animals. Why? Not because the farmer is a "mossback" or because he is slow, it is because the machinery, the mechanisation offered to him, will not suit his conditions. If we are going to bring about the benefits of the "Price Reducing System", we must give the farmer something that will reduce his costs, cut them in pieces—not on a beautiful level field or in one particular crop, like wheat, but everywhere, under all conditions, all over the world.

'I believe that the world could absorb something approaching 100 million tractors for agriculture and industry. Probably 99 per cent of the farms in the world are operated by hand or by animal power. Only a mere fraction are mechanised in the sense that a factory is mechanised. Why do factories use electricity and steam or oil engines to drive their machinery? Why don't they drive their machinery with horses or bullocks or mules? The reason is simple. The cost to the factory owner would be so excessive that he would be bankrupt in a few months.

'Why has agriculture had to continue the use of animal power? Agriculture is more important to mankind than all the other industries in the world put together—yet it is about the only industry left in the world that is still being conducted by antiquated methods. Why, gentlemen? Because until now no one has come along with machinery which will do, on all the farms of the world, everything which animals now do—and do it at a fraction of the cost.

'It is that machinery you are going to see today. We are going to show you that we can go to any country in the world, no matter what the farming conditions, and cultivate and produce its crops at half the cost at which they are now being produced.

'If the government of each country would help to inform their farmers of what this machinery will do, and if it be established on their farms, the cost of living can be brought down by half. By doing this we can eliminate the "Price Increasing System" and substitute instead the "Price Reducing System".

'I know that many of you represent countries where the farms are very small. The smaller the farm, the more urgent the need for mechanisation. There is not sufficient land to maintain the animals supplying the power and also to provide adequately for the farmer and his family. The small farmer has a life of slavery and poverty, and his purchasing power is so low that he is of little value to other industries.

'So today, we have the case of the farmer, who is the most important citizen in the whole world, still conducting his business by slavery and manual work, whereas in the factory it is done by "finger-tip control".

'This is all wrong. Agriculture should have been the first industry to be modernised and not the last.'

21 *Truths and Fallacies*

Harry Ferguson was not the first inventor to claim that the fruit of his brain and effort could have a radical effect on the wellbeing of mankind, but his arguments had a fundamental appeal. Man is doomed without his daily bread, and much of what Ferguson said struck upon the primitive awareness that dwells in even the most sophisticated of men, the awareness that in the final analysis we depend upon the land for our survival. Food is the beginning of all things, and its production, although left to peasants and faceless farmers hardly worthy of consideration, is in fact as important in today's industrialised era as ever it was. It is astonishing how this fact had disappeared from general view and how agriculture has become the poor relation among the world's industries. In many countries, the farming community is viewed with a good-natured contempt by the city dweller. And right up to the present time, governments of developing countries frequently give preference to fostering industry rather than agriculture. (It is significant that it was only in 1969 that the International Bank for Reconstruction and Development decided that in future it would give more loans for agricultural as opposed to industrial development projects.) Even today over 60 per cent of the world's population is engaged in agriculture, though it has been proved by experience in North America and Europe that efficient, modern, and mechanised agriculture will allow 5–10 per cent of the population to feed the rest employed in other industries.

Almost thirty years ago, when Ferguson pleaded for the modernisation and recognition of agriculture, his case was even better supported by circumstances than it would be today, though his argument is still valid. In other respects, however, there were serious flaws in the ideas he put forward at Bethesda to the delegates attending the International Food Conference. At the time these flaws were not as apparent as they are today, given the benefit of hindsight, and most of the delegates found themselves carried away by this assertive, quicksilver man with the emphatic gestures. They believed that truly they were witnessing something of moment; but flaws in the argument there were, and they were very grave.

Firstly, Harry Ferguson's knowledge of agriculture on a world scale was very limited indeed. Despite his claims to the contrary, this fact is evident from just one of the attitudes he adopted in his campaign, that towards draught-animals: his sweeping and vehement condemnation of them all assumed that they were indeed purely draught-animals, as were the mules and horses of Britain and North America. Had he travelled extensively outside these regions, however, he would have been forced to note that draught-animals such as cows, water buffaloes, and camels often provided the family supply of milk as well as power on the land. That milk was, and is, of fundamental importance in alleviating the already grave protein deficiencies in the developing countries, but it never entered into his calculations when he vociferated against draught-animals in general.

There were even more fundamental weaknesses in his arguments. He grossly oversimplified the issues at stake when he advocated a massive drive to mechanise the agriculture of the world. Again, had he travelled in the rural areas of Asia, Africa, and Latin America he might have been able to form a more realistic picture of the situation. Even today the leap to wholescale farm mechanisation cannot be seriously considered feasible in the developing countries of the world. The problems are immense; to begin with, most of the agriculture in these countries is made up of subsistence farming with the peasants seldom selling more than the small surplus to their own requirements. In good harvest years this surplus just might be considerable, but taken over a number of years, the annual cash income of the average peasant is derisory. Even if such a farmer were to be given a tractor and set of implements, his problems would not be over; he would need cash for fuel and repairs, and to purchase fertiliser to compensate for the loss of manure from his draught-animals if no more. Therefore, the very fabric of his existence would have to be changed from a subsistence to a cash economy, and thus the mechanisation of agriculture can only be part of a complete infrastructure of credit, and of marketing.

Another impediment in the path of mechanisation is the psychological barrier set up by considerations of risk: a peasant who is managing to feed his family at the subsistence level, but very far from plentifully, is going to have serious and well-founded fears about mechanising his holding—or for that matter adopting any other innovation—if the resultant indebtedness could possibly tip his delicate balance of survival. From subsistence to starvation is only a small step, and it is one that floods, droughts and other natural catastrophes often enough force the peasant to make without he himself imperilling his survival by adopting an innovation which for many reasons might not immediately pay off.

During another part of the Bethesda demonstrations, Ferguson had this to say about farm mechanisation:

'There is a theory that labour-saving machinery creates unemployment. This is entirely erroneous. Labour-saving machinery does the opposite. Take as an example a country where, with this light tractor, you can do as much in one hour as can be done in 100 or 200 hours by hand labour. You say that it is going to create a great deal of unemployment. No; the very reverse is true because the first great thing that happens when you save time and labour is that you reduce the cost of your product. The lower the price, the more you can sell. The more you sell, the more employment you can give.

'Some of your countries may wish to import your machines; others may be large enough to manufacture your own. If you manufacture your own tractors, you will require thousands, hundreds of thousands, or millions, as the case may be, depending on the size of your country. The moment you start manufacturing you create employment from mining right through manufacturing to transportation and distribution. To equip your factories to make this machinery will create great employment in other industries, since you must have all the machine tools, must build all the factories, etc., etc.

'There are a multitude of new jobs that will arise when you start to manufacture on a big scale. Taking it by and large, you create employment with labour-saving machinery. You don't reduce it. That is a very important thing to remember, because you will have to face it when you are telling the story of what you are going to do with labour-saving machinery on the farm.'

This is by far the most debatable aspect of all Ferguson's argument. The mechanisation of agriculture has, historically, marched in step with industrial development. As manpower has become redundant on the land it has been absorbed quickly into industry, and the increased agricultural productivity has been able to feed the city-dwelling workers. Admittedly Harry Ferguson, in the passage quoted, mentioned that the production of the tractors and implements would in themselves give employment if countries decided to undertake their own manufacture, but this alone could not take care of the many millions of agricultural workers who would be driven from the land were full-scale farm mechanisation, by some miracle, made an almost immediate reality. Industrial development, other than that of tractor and implement production, would have to play an integral part, and this issue, though extremely complex, was ignored by Ferguson.

In a country like India, where many agricultural workers can only find employment on the land for 150–180 days a year, mechanisation could bring added hardship initially, but hardship on a scale such as to make mechanisation as envisaged by Ferguson unthinkable. Usually an agriculture such as that of India and most other developing countries must be intensified—by means of inputs such as fertilisers, irrigation, and improved crop varieties and livestock, before it warrants or can benefit from mechanisation. And even then this mechanisation has to be stimulated by the squeeze of labour shortage as manpower is siphoned off for the industry that is developing alongside. Industry in turn is fuelled partly by the increasing purchasing power of the farmer as he becomes more efficient and productive. The interrelated course of events is indeed complicated and, sadly, the problems cannot be solved by the relatively simple solution of cheap and efficient farm machinery. And the idea that the 'first great thing that happens when you save time and labour is that you reduce the cost of your product' can only hold true for agriculture when there are not millions of un- or under-employed farm labourers, as there are in so many parts of the world.

I have perhaps been overharsh in pointing out the flaws in Harry Ferguson's arguments at Bethesda, for his ideas of cheap and efficient farm machinery bringing benefit to the world were true to a certain extent, but not to the extent that he believed. But his clarion cry for the recognition of agriculture in its full importance to mankind contained so much truth that it deserves a reverberant echo. Only today, with a staggering population growth on our hands and in most countries methods of agriculture that a reincarnated medieval peasant would recognise as his own, can those truths of almost four decades ago, and the farsightedness of their proclaimer, be fully appreciated.

PART TWO

22 *Back to Britain*

Fiercely British, and bristling with patriotism, Ferguson never lost sight of his ambition to have his tractors and machinery manufactured in his native isles. Perhaps in part his humble origins urged him to seek the recognition that had eluded him in Britain up to the 1939 split with David Brown and his transfer to the United States for the venture with Ford. In any case, he continually pressed Ford to enforce the same manufacturing arrangements in Dagenham, England, as those appertaining in Detroit. But as mentioned earlier, the management of the Ford Motor Company in Britain was violently opposed to any such arrangement with Ferguson. Indignant and much piqued by the rebuff, Ferguson later went to the length of writing a letter to Henry Ford to inform him that he wished to withdraw from that part of the 'gentleman's agreement' that provided for the manufacture of Ferguson System tractors by the Ford Motor Company in the Eastern Hemisphere, or more particularly at Dagenham. Henry Ford's trusted private secretary, Frank Campsell, took a horrified look at the letter and decided it would be better if it never reached its addressee. He accordingly put it in a drawer in the hope of maintaining the Ford–Ferguson relations.

Shortly after the Bethesda demonstrations, however, Trevor Knox was dispatched back to England with the task of laying the groundwork for manufacture of Ferguson System tractors and implements in Britain after the war. Such a task was certainly not easy in 1943–44, and Knox could do little but keep his ear to the ground, and look for potential opportunities. At least by this time the value of the Ferguson System was appreciated in England. The Ford–Fergusons brought in under American lease-lend, with their implements, were performing beautifully, like precision-built and jewelled-movement watches compared to the rest of the hefty and uncouth equipment in use, equipment whose direct lineage from the blacksmith's shop was plainly evident. Yet despite its miniature appearance, the Ferguson-designed equipment had heart and vigour. With deft lightness, on the level, uphill or downhill, it would break open deeply matted and compact grassland that had not felt the bite of the plough in centuries. In bracken-covered, stony and marginal land too, it worked with confident ease, and all the time the driver could control the depth of work with his fingertips. At the end of a day's ploughing or cultivating that had opened the way to putting more acres of Britain under vitally needed food crops, he could climb down from his tractor scarcely more tired than if he had spent the day driving an automobile. These advantages were recognised, and even if the Ferguson System still had its confirmed and dedicated enemies, a vast majority of users blessed it for the simplicity, ease and economy it had brought to their operations.

Trevor Knox installed himself at the Farmers' Club in Whitehall and began to send out feelers in many directions. None of these had any positive results, for as so often seems to happen in such cases, it was pure chance, luck, or fate—according to one's point of view in such matters—that eventually led to an opening. Living in the Farmers' Club at the same time as Knox was a director of the Distillers' Company. They struck up an acquaintanceship, possibly aided by the products of the above-mentioned company and by the fact that Knox did not share his employer's resistance to the odd 'wee dram'. During their conversations, the Distillers' director learnt of the reason for Knox's presence in England and of his persistent if unrewarding efforts. It happened that the Distillers' man travelled regularly to Brighton by train at week-ends and a frequent companion on the journey was an advertising executive from a company that handled publicity affairs for the Standard Motor Company of Coventry, an automobile manufacturer that was busily engaged at the time in the production of aero-engines and other war material. During the course of one train journey to Brighton, the Distillers' director told the advertising executive about Knox and his quest for a manufacturer for the Ferguson System tractor. The advertising man remarked that Standards were thinking of developing a motorised farm cart for post-war production and that if Sir John Black, managing director of Standard Motor Company, knew of Knox's quest he might be interested. From that point the message sped further along the grapevine until it reached

Sir John Black's office. Thus, by a few chance meetings and conversations, a link was established that was to have important economic consequences. For Black was immediately interested in peacetime tractor manufacture, even if at that point there were few indications of how industry would be reorganised when no longer called upon to produce war materials. It was, however, apparent, even then, that steel shortages might be restrictive for months or even years after the advent of peace.

The celebrations of V.E. day were still reverberating when Ferguson decided he must go to Britain at once, for he had been eagerly and impatiently awaiting the peace; and however good a job Knox might be doing, Ferguson never had sufficient confidence in any of his employees to allow them to work completely on their own. Like most perfectionists, he believed he could do any job better than any of his staff. Someone once defined a perfectionist as 'one who takes great pains but also gives them to others'. This was certainly true in Ferguson's case.

After a journey to Britain in a ship still fitted out for troop carrying and many hours spent in railway compartments, with no dining car or other source of food, Ferguson and his wife and daughter arrived in London appalled by the austerity that had struck their victorious land. Always a lover of cars and covetous of a Rolls-Royce or Bentley, Ferguson set among his top priorities the purchase of such a vehicle, second-hand of course as no new ones were being manufactured. He found one and felt that the acquisition helped to confirm his arrival among the ranks of successful industrialists. For Ferguson never believed, as do some, that there is merit in pretending to be less successful than he was. In fact the opposite was necessary because throughout his business life he almost invariably found himself in the role of a petitioner. His industrial success was built on the capital and plant of others, and the bigger the show of affluence and prior success he put on when attempting to persuade people to invest in his ideas, the greater his chance of succeeding. Independent industrialists are free to look like paupers if the masquerade appeals to them, but to a man like Ferguson his Rolls-Royce was a working status symbol whose function was probably more to influence prospective investors than to give its owner pleasure and a feeling of superiority. Partly for the same reasons, Ferguson always stayed in the very best hotels. (He established himself at Claridges upon arrival in London in 1945.) It must be pointed out, however, that as well as realising the advantages of displaying his wealth, the same innate good sense always steered him clear of ostentation and his display was always, therefore, in the best of taste. It seems likely that in this Maureen was his guide and mentor.

Prospects for manufacture of the tractor and implements in Britain looked bleak indeed owing to the shortages of steel that had been foreseeable even during the last year of the war. A New York consulting engineer called Abbot arrived in Britain to help Knox and together they visited plants in Sheffield, Cardiff, and Newcastle, as well as the Standard Motor Company shadow factory at Banner Lane, Coventry. This factory lay idle and forlorn, about a million square feet that had been the scene of frantic but precision toil as beleaguered Britons had worked day and night, to the ditties of 'Music While You Work', supplying aero-engines that would power the offensive against the Luftwaffe. But now the unsung factory-floor heroes, whose stolid labour had supported the heroes of the sky, had left; plant was being moved out, and the desolation was far removed from the scenes that had prevailed during Britain's 'finest hours'. Nevertheless, this was obviously a perfect factory for large-scale tractor production. The drawback was that the Standard Motor Company found themselves totally unable to make any commitment to manufacture with the steel situation as critical as it was.

Ferguson still nurtured a hope that Ulster could benefit by producing the tractor, and in July 1945 he had talks with the Minister of Finance of the Stormont government. The rest of the Cabinet were consulted and Ferguson received a letter from the government to say that in principle it was 'prepared to finance by way of loan a manufacturing plant'. But in London Ferguson, who always believed that meddling with minions was counterproductive, had taken his case straight to the President of the Board of Trade, Sir Stafford Cripps. That austere politician, who had the hapless task of steering Britain through a period as grim and bleak as any it has ever known in peacetime, and perhaps in wartime as well, found Ferguson's buoyancy and assertiveness particularly appealing. Even if the inventor had by now reached his sixties and was suffering from stress, occasional bouts of depression, and eye trouble, when he was on the upward swing he was as irresistible as ever, especially if he applied his charm in support of his arguments.

He had some compelling points to make to Cripps, quite apart from his already outlined general philosophies on the need for farm mechanisation. Britain's balance of payments was so sickly that any treatment that might improve it was of necessity attractive; therefore Ferguson promised Cripps that his equipment would be a big dollar earner, pointing out too that more home-produced food would cut down on costly imports. Thus it was in the nation's interest to sanction steel supplies for the production of tractors and implements. The appeal was one that had immediate impact, but Cripps could not commit himself at once. In the interval Ferguson decided that the time had come to organise one of his demonstrations in an attempt to tip the balance in his favour and secure the steel he needed.

The task of arranging the demonstration fell to Knox, the only Ferguson employee in England until he found a young graduate, Richard Davis, of the Ford Foundation's Institute of Agricultural Engineering at Dagenham, to help him. As in 1940, a site near what is now London Airport was chosen. Knox, Davis and another recruit to the team were forced to use tractors that were far from new, and they had to be brought into a lustrous state of cleanliness and new paint, as well as perfect functional order, to please Ferguson. In a barn on the selected farm they stripped and re-assembled, scraped, cleaned and painted for weeks on end. Fortunately some new implements had been found, still crated, at Dagenham. Ferguson came to the barn almost daily to check that the work was being properly done.

The choice of people to attend the demonstration would be crucial to its success; Sir Robert Sinclair, head of the Ministry of Production, with whom Ferguson had assiduously cultivated good relations, drew up a list of twenty or so people he thought should be invited; these included the key figures from government offices involved, from manufacturers already interested in the venture, and, comic as it seems today, a representative of the Chinese government.

The demonstration at Feltham was a very dignified occasion,

apart from one besmirching factor. Young Davis, who had helped to prepare the equipment, had been fully trained in his role of a demonstrator, with the exception that no dress rehearsal was carried out. Understandably imagining that neat jodhpurs and hacking jacket were suitable garments for demonstrating farm machinery, he turned up on the morning of the demonstration so attired. Ferguson's wrath was almost boundless.

'Let there be no mistake,' he ranted, 'a dark, single-breasted lounge suit is the correct apparel for giving a demonstration.'

Had there been anyone available to stand in for Davis, he would certainly have been banished from the field. For the question of dress among his staff assumed the proportions of a fetish with Ferguson as the years went by. No one, even when he employed several hundred people, was allowed to wear a double-breasted suit: such suits were unfunctional because you had to undo a button when sitting down and rebutton it again when getting up. Failure to attend to the button caused the suit to pull into a crease or to hang untidily. And to see an employee wearing a sports jacket and grey trousers, or a suit with too loud a pattern, would produce a peremptory order for the man to go home and dress himself properly. As for his senior executives, he frequently straightened their ties for them, or refolded their handkerchiefs, delivering contemporaneous instructions on how these operations should be carried out. Such crankiness seldom aroused the irritation that it could have, mainly because Ferguson had infused among his staff such *esprit de corps* and loyalty to the cause that they accepted the strictures and constraints in the same way as does a crack military unit. They were proud of the discipline and of the very high standards set. To have to observe such standards automatically makes an ordinary group of people into an elite corps in their own minds, and they are therefore unlikely to complain much about the restrictions imposed upon them.

As Ferguson had hoped, the Feltham demonstration of his equipment strongly influenced opinion in his favour. Sir John Black immediately saw the potential for his company in manufacturing the tractor, and the only remaining problems were those of steel and of the dollars to tool up the Banner Lane plant and import North American engines until a suitable one could be designed and manufactured in Britain.

'I need steel for two hundred tractors a day,' Ferguson told Cripps bluntly.

'I think you would be wiser to ask for steel for a hundred a day,' Sir Stafford counselled him. 'Then there'll be less risk of your being cut back.'

The gleam of battle fever lit in those blue eyes behind their rimless glasses.

'If you are going to cut me back, then I'm going to ask for material for four hundred tractors a day,' Ferguson retorted. 'The country and the world could use them, in any case, and more besides.'

Before such an assertion, Cripps backed smoothly down. Ferguson applied for material for 200 tractors a day and Cripps granted the request; in fact, not only did he grant it but he also instructed the Standard Motor Company to take on the production. This was not quite such a case of state interference in private enterprise as it might appear at first sight, for the Banner Lane factory belonged to the government and was only on lease to the company.

The question of the dollars necessary to set up manufacture was agreed at a meeting between Sir John Black, Ferguson and Sir Wilfred Eady of the Treasury at the end of September 1945. Black and Ferguson asked for half a million dollars for machine tools, five million for Continental engines, and three million for certain implement parts. The Cabinet agreed the expenditure of dollars with a minimum of fuss, though Eady in his letter advising Ferguson of government approval specifically stated that in 'any public statement on this new project there must be no reference to the amount of preliminary dollar expenditure involved without prior reference to us'.

It did not take long for Sir John Black and Harry Ferguson to have their first serious differences of opinion. The fact that such arguments arose was as unavoidable as the crash between two express trains hurtling down the same length of track but in different directions. Both men were inflexible and aggressively certain that their opinions were infallibly right. Their basic attitudes were therefore incompatible unless they happened to agree *entirely* on a given topic. They did agree in principle that they could derive mutual benefit from an arrangement to manufacture and market Ferguson System tractors. However, working out the details of that arrangement, equipping the Banner Lane factory with plant, and finalising design matters, gave rise to an abundance of minor controversies that the two men's characters quickly fanned into major issues.

Ferguson wanted, at the beginning of his association with the Standard Motor Company, to have an unwritten agreement similar to the one that governed his collaboration with Ford. It was a strange thing to want to repeat; the Ford–Ferguson handshake deal had been inspired to a large extent by the similar ideologies of the two participants. Indeed, Ferguson once referred to Ford as his 'spiritual brother'. But there was little that could be defined, even remotely, as spiritual brotherhood between Ferguson and Black. Critics of Ferguson have been, and still are, quick to give sinister interpretations to his desire for unwritten agreements, and it is of course true that it is simpler to repudiate an agreement if there is nothing in black and white. I do not believe, however, that such motives entered into Ferguson's reasoning. He had spent many years and untold energy angling for an agreement with Ford, and having finally got it why should he have wanted to leave the escape route of having nothing in writing? He could have protected himself and his patents much better with a carefully calculated written agreement; and he should have. In the later case of the Standard Motor Company, one could argue that Ferguson still nurtured hopes of an eventual link with Fords of Dagenham, and that therefore he did not want to commit to writing his accord with Sir John Black. But I consider this argument to be unacceptable and that the real motive for his desire to make 'gentlemen's agreements' was quite different: if one takes into account his constant striving for recognition, his determination to be a leader, it becomes evident that to seal agreements worth millions of dollars with nothing more than a handshake represented a high score to Ferguson in the prestige game he was constantly playing. By any standards, to persuade a man to put large amounts of capital and plant on the line without giving him any written guarantees in return was a prestigious feat, and it was one Ferguson revelled in. It was solid proof of the esteem in which he and his inventions were held.

With Sir John Black, however, and without the 'spiritual brotherhood' of his relations with Henry Ford, a crisis arose as soon as 1946 when the Banner Lane factory was still being equipped for production. Roger Kyes was sent for; he dropped his Detroit problems to travel to Broadway in the Cotswolds where the Fergusons were living in a hotel called Farncombe House. He arrived late at night and settled into the Lygon Arms, but he was urgently called to the Ferguson's hotel early the next morning. He arrived at the house to find Maureen and Betty tearful and distraught and to be greeted by the news that Harry Ferguson was ill in bed. The doctor in attendance confirmed that he was very weak. Kyes asked whether everything possible had been done and Maureen said there seemed little hope for him.

'In that case,' said Kyes, 'would you mind if I went to talk to him. I don't think I can do him any harm.'

Kyes went upstairs. 'I knew goddam well something had scared the living hell out of him as soon as I saw him,' he remembered.

Ferguson was inert, in a deep state of depression, and Kyes talked to him for about three hours. It emerged that Black was going to withdraw from his agreement to manufacture the tractor as a result of a row that he had had with Ferguson. Kyes, who of course knew Ferguson very well, talked ceaselessly and in a hopeful vein about patching up the quarrel, and suddenly Ferguson sat up in bed. 'I want a hard-boiled egg,' he announced.

Kyes arranged to have lunch with Sir John Black a few days later and during the course of their conversation he was able to pour oil on the troubled waters and suggest that a proper written agreement should be drawn up between the Ferguson organisation and Standard Motor Company. Black was agreeable, so Kyes and John Turner drew up the terms of a ten-year contract that was fair to both parties, and which Sir John Black agreed to sign. Ferguson was still ill in bed when Kyes arrived to tell him of the arrangements he had made. When he heard the news he was highly elated, well nigh leapt out of bed, and regained his old jauntiness almost at once. Within a week he had regained all his contentiousness too. 'He was just tootin' around giving everybody hell about some design,' Kyes remarks. 'And he was so mad at me for having made that agreement he gave me hell too.'

Thereafter the production plans went relatively smoothly, and the first of many hundreds of thousands of Ferguson System tractors rolled off the Banner Lane production line in the last weeks of 1946. The tractor, which was almost identical to the one being built in Detroit by Ford (apart from having a four speed gearbox and slightly more powerful engine) was the famous TE 20 model, the well-loved 'Fergie' of farmers in countries all over the world.

23 *The Break*

Events at the Ford Motor Company in Detroit took some strange turns during the war and in the years immediately following it. That such chaos could have existed within the company almost defies belief; and there was brutality and evil too. Matters degenerated particularly after 1943 when Edsel Ford died. He had had, in quick succession, an ulcer, undulant fever and cancer, all of which he bore with dignity and fortitude. His father, who had been hostile during his illness, was grief stricken; he had refused to believe that Edsel was ill, despite his deathly appearance, insisting that nothing ailed him that his chiropractitioner could not cure. Five days after his son's death, Henry Ford at the age of eighty reassumed the presidency of the company. Edsel's son Henry Ford II, aged twenty-six, was called back from the Navy to help run it. However, the company was really being run by Harry Bennet, an ex-sailor and pugilist who had worked his way into an extraordinary position of favour with Henry Ford Senior. Bennet was a bully and a gangster. He surrounded himself with a group of armed thugs who perpetuated appalling acts of brutality in the Rouge plant, and in the workers' houses too. The tentacles of Bennet's Machiavellian rule reached everywhere; his spies observed everyone. When Bennet wanted to do something to which the other directors did not agree, he would go away in his car and return a few hours later to say that Mr Ford agreed with him and wanted it done his way. Since Bennet was the access path to the old man's ear, no one could dispute what he reported Ford as saying. It later emerged that sometimes he had not been to see Ford at all. Finally, after Sorenson had been fired over the telephone while he was in Florida, and his personal car removed by the local Ford agent—and this after forty years of service—the power struggle became one between Harry Bennet and Henry Ford II. With the aid of his mother, and Clara Ford too, the young man manœuvred his frail old grandfather into naming him president of the company and giving him a free hand to run it. But he only gained effective control from Bennet after some tense meetings during which he felt it necessary to carry a gun.

As the war drew to a close and military contracts were curtailed, Henry Ford II found himself in control of a bleeding colossus that was staggering about without sense of direction, in danger of collapse because it was losing blood at the rate of almost $10 million a month. These heavy losses stretched through the latter part of 1945 and the first half of 1946, and it should be explained that it was not the termination of the war alone that put the Ford company into such a plight: equally responsible was the abysmal level of organisation and administration within the company. Henry Ford I had never been interested in such refinements as detailed cost-accounting, and he never thought it worth while to employ people who were. He ran the business like a village store and the sum in the company's bank account at the end of the year was its 'profit'. He even fired the few accountants he had on one occasion when his son Edsel had wanted to provide them with more office space. The company was run largely by hunch. From 1900 to 1946, no outside accountants ever audited the Ford books, and the company did not really know what operations were profitable. Indeed so badly was it run that it steadily lost its position of pre-eminence in the automotive world to better organised competitors such as General Motors. In 1946, therefore, it needed some modern management techniques, and it needed them quickly. Henry Ford II had to find high calibre executives and get rid of many of the old guard, and so after acceding to power he fired about a thousand major and minor supervisory staff, an event that caused sufficient stir for *Fortune* magazine to comment on it.

One of the first executives to join the Ford company, as a rescuing fireman, was Ernest Breech from the Bendix Aviation Corporation where he had been president. Breech, an astute self-made man, closely examined the tractor manufacturing operation and decided that it could not go on in its present form. Indeed it was true that Fords had made a consistent loss on the production of the tractor for Ferguson. They claimed that the operation cost them a total, over the years, of about $20 million. They seemed to hold Ferguson morally responsible in some way for this loss, but it was in fact largely caused by the strictures of the Office of Price Administration; for O.P.A. would in principle allow no price rises during the war. Fords did manage to obtain one of $60 on the selling price of the tractor to Harry Ferguson Incorporated, but Ferguson only passed on about half of the increase to the farmer, so neither was he making much money on the tractor. (His company's financial success was founded mainly on the sale of implements.)

Apart from the financial aspects, friction between the Ford and Ferguson companies had been growing towards the end of the war. This was epitomised in an incident early in June 1944 when a Ford executive telephoned his administrative counterpart in the Ferguson company.

'How long will it take you to get out of B building?' he asked curtly. 'We've got a new government job and you Ferguson people will have to go to the Highland Park plant immediately.'

'I'll have to find out from the powers-that-be in the Ferguson organisation whether or not the move can be immediate,' replied the Ferguson executive.

'Yes or no, hell! If you don't get out immediately you'll find yourself in the street,' came snapping down the line.

Such incidents could only add to the fundamental feeling of uncertainty that Ferguson staff inevitably had towards the Ford company. Their own organisation was dwarfed by its colossal supplier, and the relationship could only be looked upon as one of grace and favour at the best of times. From the very beginning, a number of people believed that it could not last indefinitely. The well-known attitude of Henry Ford the elder towards patents probably contributed to that belief. He had made an open stand against the patent system, and during a court case in 1928 his patent lawyer declared, 'There is no power on earth, this side of the Supreme Court of the U.S., which could make Henry Ford sign a licence agreement or pay a royalty.' In fact, the biggest suprise to many people was that Ford had ever made his agreement with Ferguson at all.

In early 1946, a Ford representative suggested that perhaps the Ford family should have about a 30 per cent interest in the Ferguson company. Then Ernest Kanzler, Henry Ford II's uncle and adviser, suggested that it might be better for the Ford Motor Company to acquire a controlling interest in any new tractor distribution company that might be established. Kyes knew full well that Ferguson would not agree to such an arrangement, and he said so. By July 1946, Ernest Breech had arrived at the Ford Company and at a meeting with Kyes and another top Ferguson executive, Horace D'Angelo, he stated that Ford wanted 51 per cent of a new company. (Kyes maintained that while Breech was still with Bendix, he had tried to persuade Ferguson and Kyes to come to a 51–49 arrangement under which Bendix would manufacture the tractor. Who first approached whom for these unsuccessful talks later became a point of some importance.)

The idea that the Ford company should take a majority share in the Ferguson business aroused indignation. Kyes and Ferguson did feel that members of the Ford *family* had a right to an interest in the Ferguson business, in view of Henry Ford Senior's role in launching it; but the idea that the Ford Company, and in particular its executives, should have a share aroused the Ferguson directors to a white fury. What is more, the Ferguson directors began to suspect that there was an insidious plan behind the Ford attempt to take over their company, a plan that involved more than just putting Ford back on the path to profitability. It seemed to them that creating a new distribution company with a Ford majority holding would provide a very convenient means of offering inducements to attract the outstanding executives Ford needed. High salaries alone were unattractive because of the 85 per cent income tax they could bear; but capital gains on share holdings were taxed at only 25 per cent, and no high calibre man would be likely to accept a Ford offer without being given the chance of accruing capital. In this respect, the Ford Company was in difficulty because its own share capital was vested mainly in the Ford Foundation and the Ford family: to have a new company in which it could offer shares to top executives would be more than useful, the Ferguson directors reasoned. And their shadowy suspicions took on substance when, in early November 1946, there was another meeting between the Ferguson and Ford executives at which Breech declared that Ford wanted 70 per cent of the equity of a new company that would *not* include the name Ferguson. Of that 70 per cent, 40 per cent would go to the

Ford Motor Company and 30 per cent to the Ford executives in person. The 30 per cent that was to be granted to the Ferguson company would be all that it would get; there would be no royalties on Ferguson patents or allowances for goodwill.

This bombshell was delivered in the thinly veiled terms of an ultimatum and, sick with anger, the Ferguson executives could do nothing but leave the meeting. A few days later, the Ford Motor Company advised Kyes that as from December 31st it would no longer produce tractors under the existing agreement. They did provisionally agree, however, to continue producing tractors for the Ferguson company to market for the six-month period ending June 1947. Then in November 1946, the Ford Motor Company established a new tractor distribution company called Dearborn Motor Corporation, a company which in effect was a substitute for Harry Ferguson Incorporated.

The blow to Harry Ferguson Incorporated of the loss of their tractor supplier was grave to the point of disaster. It happened about the same time that Ferguson himself had his first serious row with Sir John Black in England and when he almost saw that venture founder before it was launched. It was scarcely surprising, therefore, that Ferguson was in such a sorry state when Kyes reached his home at Broadway, as mentioned in the previous chapter. Had both his American and British sources of manufacture fallen through at the same time, he might as well have died, unless he gave up his life's dreams, a renunciation which would have been tantamount to death for a man like Ferguson.

But once Ferguson was 'tootin' around giving everybody hell' again he was as angry and bristly as a fighting cock squaring up for battle. Before he could wage war on the Ford and Dearborn companies, he and his North American staff had to take urgent steps to find another manufacturer for the tractor. The abyss of complete ruin yawned before the company. Its only strong point was that it had cash reserves, about $6 million, but even these provided a problem. Should the company continue to place orders with its suppliers for implements on the assumption of being able to sell them, with or without a tractor? Or should as much of the cash as possible be conserved? If they cut down too far on their implement orders, the suppliers would look towards other outlets, and in particular towards Dearborn Motors. And the

Ferguson company's distribution and dealer network too was on the horns of the dilemma. Would Ferguson stay in business or not? If not, the various companies owed it to themselves to link up with Dearborn. Not surprisingly the Ferguson network began to come apart at the seams. No one can blame the companies for looking to their own interests in that period of uncertainty and rumour; and Dearborn Motors' statements did absolutely nothing to give the impression that Ferguson had any future. According to them it was to be all Ford and Dearborn from July 1947, and many distributors were approached and offered new franchises.

Somewhat hastily, and perhaps in some desperation, Kyes arranged to purchase a large empty plant in Cleveland, Ohio, for $1·9 million with the intention of tooling up for tractor production. In order to inspire confidence and convince people that indeed Ferguson was staying in business, a distributors' meeting was immediately held at the Cleveland plant; but it was going to require more capital than the Ferguson Company had available to run the first tractor out of that derelict factory. An extra $8 million were necessary, and an underwriting agreement for public financing was sighed with Eberstadt and Company of New York. Initially, it seemed that the funds would not be difficult to raise. Business experts who visited the Ferguson management came away praising the ability, spirit and exceptional qualities of the team. These qualities, they considered, could lift the company out of almost any difficulty, including the monumental one then facing it. However, it was not easy to persuade potential investors that Ferguson could indeed be reborn and compete against the hostile might of the Ford and Dearborn Motor Companies. These companies made their enmity evident in a subtle way, but not so subtly that the implications escaped those with an interest in the convoluted affairs of Ford and Ferguson.

Two other factors contributed to the difficulty of raising the $8 million. Firstly there was a decline of stock prices on Wall Street, and secondly a strike severely curtailed telephone communications during a critical period. The financing had to be withdrawn, and at that point the hitherto cohesive and enthusiastic Ferguson management began falling into disarray. Roger Kyes's dynamic seemed, to the rest of the board, to degenerate into a sort of fiddling while the flames of the crisis rapidly consumed all chance of rescuing the Ferguson operation. The Cleveland plant was sold again, at a slight profit, and the company's liquid position was therefore again very strong. Many of Kyes's colleagues believed that his only real interest was in winding up the company and realising the very considerable sum of money that would have accrued to him from his shares, for Harry Ferguson had given him a large holding. This certainly was what Ferguson came to believe and he viewed it as a treacherous betrayal. He vituperatively referred to Kyes as being 'mean' and 'yellow' in his unwillingness to fight on. These were harsh words to use in describing a man who had done so much to build up the Ferguson company, even if the basis for the accusation was true. Kyes recalled that his policy was to keep his powder dry and not to disperse the assets of the company in an inventory of implements and parts, hence his reluctance to make commitments, and his apparent *laissez-faire*. It would not be fair to pass judgement at such a distance of time, but the consensus of opinion is that Kyes had indeed given up the idea of re-establishing the company. Despite this, the Ferguson management contacted almost every company in North America that might possibly be in a position to take on the manufacture. The list of companies reads like a 'Who's Who' of American business, but ultimately the concerns showing genuine interest were whittled down to a very small handful, among them Willys' Overland, Kaiser-Frazer and General Motors.

Meanwhile, the Ford Motor Company was preparing to launch its new tractor in July 1947, and it did so with considerable éclat. To the surprise of some, though not of 'informed Detroiters', the new Ford tractor, the 8N, incorporated an unmodified Ferguson System of hydraulics and linkage. It was in fact an improved version of the earlier Ford/Ferguson, incorporating a total of twenty-two modifications. The most important of these was a four-speed instead of a three-speed transmission; the others were minor and ranged down to the colour scheme. The Ferguson engineers claimed that even the four-speed transmission was an improvement that they had developed prior to the break with Ford, a claim borne out by the four-speed transmission in the Coventry-built TE 20 model.

Time magazine of July 21st, 1947, described the ceremony to launch the new Ford tractor. Three hundred guests 'quaffed beer and cocktails, munched cold meats and salads buffet-style then watched a new Ford tractor plough the hard clay of the soil outside. Said young Henry . . . "Since the days of my grandfather we have always had one foot in the soil and one foot in industry. We will continue that policy." '

'The foot which Henry put in the soil last week also came down hard on the neck of an angular Irish inventor named Harry Ferguson.' The article continued to describe the unique arrangement which had bound the Ford and Ferguson interests and outlined how Ernest Breech had begun to straighten out the Ford Motor Company's 'muddled accounting system', and had ended the tractor contract with Ferguson.

Ernie Breech also had a personal interest in tractors, *Time* stated. Henry II had lured high-priced men like Breech into the company by giving them stock in a new farm equipment company, Dearborn Motor Corporation. Thus the personal fortunes of top Ford officials depended on Ford's ability to make and sell a tractor of its own. Many of the Ferguson dealers had deserted to Dearborn Motors. They would find their new product familiar. At the party on Dearborn Motors' experimental farm—purchased the previous year from Henry II—those who saw the new tractor thought it looked so much like the Ford-Ferguson machine that many predicted a patent squabble, *Time* reported.

The term 'squabble' as used to describe the battle that was brewing was tantamount to describing the Vietnam War as a skirmish. By putting a Ferguson System tractor on the market and selling it through Dearborn Motors, Ford administered a potential *coup de grâce* to the already faltering Ferguson company. But Ferguson was not the type to see his business ruined without fighting every inch of the way.

At the time of these momentous events, Ferguson was himself in a clinic in Zürich for treatment of some sort of 'chronic toxaemia' as he himself termed it, adding that it was a complaint that had dogged him for years. That his health should have declined to a low level during that period of stress and concern for the future was surely no coincidence. However, the meticulous attention of a Swiss clinic, and the professional concern of highly-paid private doctors, soon had Ferguson feeling better. By

November 1947, he was well enough to carry out a trip to the United States to see to things himself. By this time he had lost faith in Kyes, a process that had been accelerated by the reports on Kyes's performance that he solicited from other members of his management. This means of obtaining information was both unethical and unpleasant, but Ferguson would have answered an accusation to this effect by stating haughtily that the ends justified the means.

He went to New York with John Turner, his financial adviser. When he arrived he was in a 'quarrely mood'. This was more than a pity: for Kyes and one of his fellow directors, Horace D'Angelo, had succeeded in interesting Charles Wilson and other executives of General Motors in tractor manufacture. He had organised a meeting between them and Ferguson. During the course of the meeting, a General Motors man, by the name of Evans, made a comment on the size of tractor necessary for North American farms. Ferguson disagreed and did so in such an intractable manner that the rest of the meeting was taken up with the ensuing argument on tractor design. Although General Motors probably had many other reasons for ultimately not deciding to manufacture Ferguson tractors, their inventor on that occasion can have done nothing to help his cause. The Willys' Overland company got as far as attending a demonstration that Ferguson himself organised, but they were only prepared to come to an agreement if they had a controlling interest in the new company.

While these negotiations were in progress, relations between Ferguson and Kyes were deteriorating further. Since it was a Ferguson characteristic that once he had lost faith in someone he had absolutely no further use for them, he did not wait long before holding a board meeting at which, in his own words cabled to England, he 'made (a) fierce attack (on) Kyes for his illdoings and scared him into a good settlement'. That board meeting was thoroughly unpleasant for all concerned, except perhaps for Ferguson who, rightly or wrongly, was bitterly enraged and therefore had the satisfaction of wreaking vengeance. Kyes on the other hand, even if he had to tolerate a withering volley of abuse, had the satisfaction of having his holding in the company bought out for $1·25 million.

Harry Ferguson's spirited belligerence was directed in even

greater measure against the Ford and Dearborn companies. The services of a large New York firm of lawyers, Cahill, Gordon, Zachry and Reindel were engaged to prepare a complaint against the above mentioned companies and their directors. A partner in the firm, James A. Fowler, held long sessions with the Ferguson management, which, after the demise of Roger Kyes, was led by Horace D'Angelo. D'Angelo had previously been the company secretary. He was an accountant who had started with very few of life's advantages; he had begun work at the age of twelve and by sheer diligence and intelligence reached the high echelons of industry. A small sandy man with crinkly hair and freckles, he shared some of the indomitable spirit of his employer, and he therefore entered energetically and willingly into the battle against Ford. The die was cast on December 11th, 1947, in the Gotham Hotel, New York, when the following was hand-written on a piece of hotel notepaper, signed by James Fowler and D'Angelo and approved by Harry Ferguson.

'It is our understanding that you do not wish us to entertain any suggestion of a settlement on the basis of a lump sum cash payment, however large, or royalty payments for a licence, or a combination of the two. Unless they are prepared to pay the full amount as set forth in the complaint, the only way the suit can be settled is through a friendly meeting between Henry Ford II and yourself. Unless Henry Ford arranges to meet with you, we understand that you desire that the complaint be filed and the suit prosecuted with a view to the protection and enforcement of public interest involved, as well as the private rights of yourself and your company.'

The 'full amount as set forth in the complaint' was no less than $251,100,000 plus 'reasonable attorneys' fees and costs'. When Breech, leafing through the first copy of the complaint he had seen, came across this staggering figure, he was dumbfounded. 'My God! The Marshall Plan,' he finally found the wit to say. That the Ford and Dearborn executives had expected a *patent* suit was evident from correspondence with Dearborn Motors distributors; but instead they found themselves with a mammoth anti-trust suit on their hands. It was both an embarrassment and no small source of worry for the organisation, even for an organisation as large as Ford.

Throwing down the legal gauntlet rebuilt the morale of the Ferguson team almost overnight; they were crusading again, they were fighting for the rights of their small company in the face of a Goliath of industry. At the same time, the announcement of the lawsuit lifted much of the doubt that had been hanging over the ability of Ferguson to re-establish his company. His position was made abundantly clear by the complaint as lodged. Under a heading 'The Conspiracy', the following clauses figured:

> To carry out the conspiracy and to achieve the objective, (the) defendants agreed to adopt and did adopt the following campaign and plan of action:
>
> (*a*) Defendants would deliberately copy the Ferguson System tractor and line of implements, infringe the Ferguson patents without regard to their validity, and unlawfully seize and appropriate the inventions, engineering developments, designs and ideas and information communicated to Ford Motor Company under the agreement of trust and confidence.
>
> (*b*) Defendants would destroy the Ferguson Company's sources of supply of implements and would induce or compel its implement manufacturers to break their contracts and sever their connections with the Ferguson Company.
>
> (*c*) Defendants would take over in its entirety the distributor-dealer organization built up by the Ferguson Company.
>
> (*d*) Defendants would exclude the Ferguson Company from access to the market by imposing upon Dearborn distributors and dealers, and even upon Ford car dealers, the condition, agreement or understanding that they would not deal in tractors or implements sold by the Ferguson Company, and by preventing other distributors and dealers from dealing in tractors or implements marketed by the Ferguson Company.
>
> (*e*) Defendants would impede and frustrate the efforts of

> the Ferguson Company to arrange for the manufacture of its tractor and to provide capital funds required by reason of the break with Ford Motor Company.
>
> Defendants have aggressively and coercively carried out such campaign and plan of action . . . and have in large part accomplished their objective.

The complaint continued for another twenty-five pages to detail the manner in which, according to the Ferguson Company, Ford and Dearborn had carried out the conspiracy. The accusations were manifold: enticing key Ferguson employees to work for Dearborn Motors; making press announcements calculated to destroy the business of the Ferguson Company by persuading distributors and dealers to break their connections with it; buying out key implement manufacturers (such as Woods Brothers of Des Moines) who for many years had been subcontractors to Ferguson, and so on. Under the American Antitrust laws, Ferguson was able to claim threefold the damages his company had incurred. Since the company's business, deemed ruined, was valued at $80 million, this accounted for $240 million of the total claim. Patent infringement on the 37,000 tractors already built and sold by Dearborn Motors accounted for the other $11,100,000 claimed. It is significant, incidentally, that Ferguson employed a firm of public relations consultants to make maximum capital out of the launching of the case.

The Ford and Dearborn companies refuted every allegation, claimed that the Ferguson patents were invalid, the press announcements purely coincidental, and that, in Henry Ford II's words:

'The mere suggestion that there has been "conspiracy" or "unfair competition" in this situation is ridiculous.

'The blunt truth about this relationship is that it made Mr Ferguson a multi-millionaire and cost Ford Motor Company $25,000,000 in the process.

'From its very beginning Ford Motor Company has tried continuously to find new and better ways of making farming easier and more profitable through the use of machines. We started making farm tractors in 1917. From 1939 to the middle of 1947, our tractors were distributed through Harry Ferguson,

Inc. With our help, the Ferguson Company recruited a sales organisation from among Ford dealers.

'Our arrangement with the Ferguson people was terminable at will by either party.

'By 1946, this arrangement had become intolerable.

'We hoped that we might find a more satisfactory way for continuing the relationship with the Ferguson Company. We discussed a number of proposals with them. All of our proposals were rejected.

'The only choice left to us was to break off the relationship. We did so the latter part of 1946. But—although we did not have to do it—we continued to supply Ford tractors to the Ferguson people for six months more so that they could have time to establish other manufacturing sources.

'Having failed to do so, Harry Ferguson and his company now apparently hope to recover from Ford Motor Company for their failure.

'Mr Ferguson always was—and still is—free to make other arrangements for the production of tractors in this country—provided, of course, that he does not continue to trade on the Ford name.

'The complaint is full of untruths. Many important facts are omitted, and others have been distorted and twisted out of their true meaning. We will be very happy to meet all of the allegations in the complaint at the proper time and place.'

It would be tedious to enumerate all the legal details of the suit that, from filing in January 1948, continued until April 1952. Only an outline of the lengthy litigation will be given here, but there are some interesting human aspects of the case that are worthy of note. The first of these concerned Willie Sands who, it will be remembered, was the staunch and gritty engineer responsible for so much of the development work that went into creating the Ferguson System. Subsequent to his brief visit to Detroit to help finalise the design of the Ford/Ferguson, he had stayed in his beloved Ulster where he tinkered with farm implement ideas and kept in regular touch with Harry Ferguson, John Chambers, and the other engineers in Detroit. He was in effect acting much as a consultant, though Ferguson had conferred upon him the title of Chief Engineer. That Ferguson would have preferred to have Sands working closely with him he made very clear, but Sands was stubbornly attached to Ulster, to a degree that he would never live anywhere else. Despite this, Ferguson treated him reasonably well in terms of salary and bonuses, paying him a total of £19,000 ($76,000) through the years 1939–45. On the other hand, it could be argued that one who had done so much to make the Ferguson System a viable proposition should have shared more fully in the spoils once the invention became the basis for a highly lucrative enterprise. This was in his heart what Sands felt; and Ferguson indeed recognised how much he owed to him. Nevertheless, Ferguson was irritated beyond measure by Sand's pig-headed refusal to 'stay on the job' after 1939, and he allowed this to cloud the issue of what he already owed Sands for services rendered during the years from 1911 to 1939. There was in Ferguson's attitude a certain pique at the independence of his wayward employee, even though at the same time he held him in high esteem and affection. Thus Sands, and Greer too, had the galling experience of seeing men like Kyes taken to the Ferguson bosom and made into millionaires. (And the Sherman brothers were also bought out for large sums of money.) Even when it was not a question of money but of simple prestige, they found it irksome that John Chambers and the American engineers apparently counted for more than they did. Such men were mere *parvenus* in their opinion.

In the autumn of 1947, shortly after the break with Ford, Sands wrote Ferguson a letter that was heavy with discontent. 'I always seem to be living for the future. It is like the notice I once saw in a pub: "Free drinks tomorrow". One thing that plagues me is my long memory.'

Ferguson's reply was sharp, and Sands was suspended from the company for a while. But by February 1948 amends had been made and Sands was being re-promised some preference shares (in Harry Ferguson, Inc.) that had been under discussion since shortly after the war. He was also allowed to purchase £300 of shares at par value in the Coventry-based company. All seemed restored to normal, but in fact Sands was deeply grieved over another matter: Ferguson did not think to tell him of the filing of the lawsuit and Sands read of it by chance in *Newsweek*.

This was the drop that finally made his cup of misery overflow.

Ford Motor Company became aware of his disenchantment and in February 1948 he received a trans-Atlantic telephone call from Harold Willey. Willey, it will be remembered, had worked for Ferguson in the Huddersfield days and had gone to Detroit with the Ferguson team in 1939. Subsequently he left Ferguson to grow carrots in Texas, but by the time the Ferguson lawyers lodged their complaint against Ford and Dearborn, he was working for Ford. In Sands's state of deep-seated resentment against Ferguson, it was not difficult to persuade him to testify along lines that were basically favourable to Ford. Sands, thinking back over the many years during which he had worked closely with Ferguson on implement and tractor design, often solving problems with simple though brilliant solutions, felt that he had not been granted the recognition he deserved. He could not see how his reluctance to live anywhere than in Ulster affected the issue: his contribution to the Ferguson System was undeniable and hence reward should have been forthcoming. He misunderstood one vital factor, however, in his working relations with Harry Ferguson. The subtle, razor-edge distinction between invention and design eluded him completely. Where, in fact, does the work of the inventor finish and that of the designer begin? That Sands was in his way a genius is not in dispute, but really he only worked within guidelines established by Harry Ferguson. Once that overall aim had been defined by Ferguson, Sands often provided brilliant solutions to its implementation through his detailed design work. But the fact has to be faced that without the Ferguson stimulus (that is to say when Ferguson was in Detroit or Coventry, and Sands was in Belfast), Sands never developed any original ideas of his own.

Sands believed, or was led to believe in his state of discontent, that he was the true inventor of the Ferguson System. He suggested to Harold Willey that someone from the Ford Motor Company at Dagenham should make contact with him, and subsequently he and Patrick Hennessey, now Sir Patrick, met in a Dublin hotel. He offered to tell 'the story' if Fords would cover the expenses of his trip to the United States for him to deposit his evidence. This Fords were more than willing to do, for if it could have been proven that Sands was the true inventor of the Ferguson System, the patents would have been invalid. Thus Sands entered the battle in the Ford camp.

The complaint was filed by the Ferguson lawyers in the Southern District of New York. There was no problem in serving it on the defendant corporations, but under Federal law individual defendants could only be served while in New York. Since all of them were normally resident in Michigan, something of a cloak-and-dagger situation developed in order to pounce whenever a defendant came to New York.

Henry Ford II had his complaint served upon him as questioners were thronging around him after he had given a lecture in New York, but he smiled unconcernedly. Other defendants received theirs as they alighted from trains at Grand Central Station, the lawyers having been tipped off as to their time of arrival. But the main battle in the early stages, and one which cropped up again later, was that of venue. Fords wanted the case to be heard in Detroit; Ferguson and his lawyers would not hear of a change, claiming that the vast numbers of people in Detroit with direct or indirect connections with the Ford Motor Company would make it impossible to find an unbiased jury. On each of the other three occasions that the matter of venue was raised, the last time before the U.S. Supreme Court, Ferguson won a favourable decision to keep the case in New York.

In March 1948 an incident worthy of comic opera occurred in London. Henry Ford II had come to Britain on business. In his own words, as related to the press, 'I was going into Claridges and saw a car outside. I thought to myself "that's a beautiful car" when Mr Ferguson stepped out. He said "How are you Henry?" Neither of us knew what else to say. We both got into the elevator and we had some stilted conversation for three floors. I got out and he got out. He said, "I'm going to my room. I always have 301." I said, "You're not. That room is mine." So we parted.'

The way Ferguson later told the story tallied exactly apart from its ending. For in his version, as they parted, he snapped, 'All right, Henry. You can have my room, but not my System!'

The deposition of evidence for such a case was a mammoth task. It began in June 1948 with the examination of key Ferguson employees such as D'Angelo and his fellow directors. But these examinations were suspended when Willie Sands arrived in New York aboard the *Queen Mary* and gave his address as care of a senior Ford Motor Company executive, thus positively identifying himself as hostile to the Ferguson cause. The Ferguson lawyers immediately searched for him and, having located him at the Ritz-Carlton, issued a subpoena so that they could examine him as an adverse witness. They wanted to do so before the Ford patent lawyers spent too much time with him.

Poor Sands had dropped himself into a lions' den. The first day or so he was 'flippant' under legal questioning, but for high calibre lawyers he was easy to trip up. By the end of the third day he was much subdued. He had made the mistake of coming in with a briefcase one day and the Ferguson lawyers pounced, resolving to get every document out of it. 'Had you been representing him, it would have made you sick to your stomach,' one of them remarked. And the lawyers wrote to Ferguson saying, 'We have succeeded in destroying Sands's usefulness as a witness against us. He is already a liability to our adversaries and we are considering the possibility of taking steps that would make him a serious liability to them.' This was an overstatement of the situation; nevertheless the reasons for Sands's defection to Ford were, by the very nature of their subjectivity, of negative value in supporting a legal position even if they were understandable on the human level. After Sands had been examined and all the circumstances of his presence in New York exposed, he went to Dearborn for several weeks. This caused Harry Ferguson great concern because, as he said, 'Sands knows a lot more about the Ferguson System than anybody Ford has'. Fortunately for Ferguson, Sands's usual nostalgia for Ulster intervened and he went home. Ferguson had been trying to persuade his lawyers to sue Sands for large damages in respect of blueprints and documents he had turned over to Ford. He believed the shock of such an action would smoke Sands out of Dearborn, but in the event such an action was unnecessary—and the lawyers were against it in any case.

Ferguson was continually scheming as to ways of intimidating Ford and Dearborn executives to the point that they would cry out for peace and a good settlement. He believed that Patrick

Hennessey, and perhaps Henry Ford himself, should be charged with bribing Willie Sands, gleefully wondering just what the effect of the 'fear of a term in gaol would have upon them' and upon other Ford executives. The Ferguson lawyers were continually having to curb his aggressive impulses to launch further suits in all directions. And they had an even harder task in keeping him on the straight and narrow when the time came for him to deposit his own evidence in September 1948.

Ferguson had happily believed that he could give this pre-trial testimony in one week and was appalled when the lawyers wrote to tell him that it would require about a month. It turned out that he was under examination for no less than three months; and they were three electrifying months for all concerned. Ferguson had taken his wife and daughter to New York with him and they set up quarters at the Gotham, a hotel that was a comfortable anachronism in New York. (Only its elevators were not calm; their ricketiness not only offended Ferguson as an engineer, but also scared him nearly witless.) Although he was at first worried lest his wife and daughter be called for examination, their presence in New York was certainly the mainstay that allowed him to come through the ordeal with the sustained pugnaciousness of a terrier. Ordeal is certainly not too strong a word to describe those three months—for all concerned. Ferguson saw his deposition of evidence not as such but as a court room proceeding involving verbal tilting and point scoring. He could not resist snapping out a biting reply when offered an opening to make his examiner appear foolish. Nor could he resist volunteering information, to the harassment of his own lawyers who continually tried to curb him.

The main load fell upon a lawyer called John Sonnet, who had left his post as head of the Anti-Trust Division of the Department of Justice to return to his firm and take on the Ferguson case. The amiable and brilliant Sonnet, who became a close and devoted friend of Harry Ferguson, found the *enfant terrible* performance of his client very troublesome.

On one occasion, when Ferguson was giving one of his lengthy replies and gratuitously offering information, he said, 'Now, to illustrate my point, I have a document here that I think you should examine.' And he extracted a paper from his briefcase and passed it to the examining lawyers. John Sonnet was alarmed, to say the least, for neither he nor any of his colleagues had seen the paper; and they were hardly surprised when the examining lawyers, finding themselves with a new trail to explore, bore down vigorously upon Ferguson.

'Now, Mr Ferguson,' Sonnet told him at lunch, 'you must never do that again. I want your word of honour that you will never take anything out of that briefcase and produce it without our having been over it and knowing what it is.'

'All right,' replied Ferguson, 'I'll give you my word I won't do that again.'

But the very next day, he said to the lawyer who was examining him, 'You really don't understand what this is all about. You're a very ill-informed man and to enlighten you, I have a piece of paper in my pocket.' And he produced it. Afterwards Sonnet remonstrated with him indignantly.

'You gave me your word that you wouldn't produce papers like that out of your briefcase', he said in hurt tones.

'That's right', agreed Ferguson, 'but I took it out of my pocket.'

Ferguson's antagonism was also a source of great difficulty at times. Indeed it led to a precedent in American law proceedings in the following way. The principal lawyer for the defendants in the case was Whitney North Seymour, a distinguished leader of the American Bar and a very courteous man. Ferguson continually took violent issue with him, stepping well outside the conventional conduct for such procedures. Within only a few days of beginning his deposition, Ferguson was being so hostile that his lawyers were concerned as to the impression his testimony would make in the record. He *was* being provoked to a great extent, and one way to justify Ferguson's fulminations was to amplify the extent of that provocation. Frequently, therefore, Sonnet accused Seymour of shouting at the witness and of ballyragging him. Seymour and the other Ford lawyers also had their difficulties compounded by the question of smoking. As a non-smoker of cigarettes, Ferguson refused to allow anyone to smoke them in the room where depositions were being taken. His own lawyers soon accepted the deprivation, for after all the injunction was brought by a client paying handsome fees. But the

Ford lawyers bore the cross less easily and whenever, in a moment of stress, Seymour reached for a cigarette and began to light it, Ferguson, with the gleam of battle in his eyes, shouted peremptorily, 'No, no! Mr Seymour!'

The examination of Ferguson became ever more animated until, after about three weeks and much mutual provocation, Ferguson told Seymour that his question was a 'piddling' one, that he, Seymour, was a 'gangster' and that he would love to be head of Ford Motor Company so that he could fire him for stupidity. Sonnet at once had to make the usual speech for the record stressing the provocation that was being meted out to Ferguson. While he was leaning forward earnestly to make this point, Ferguson, out of his field of vision, was thumbing his nose angrily at Seymour. Another lawyer nudged Sonnet who at once caught the gist of what was going on and declared a recess. The Ferguson lawyers immediately contacted the judge responsible for the case to request a meeting with him and the Ford lawyers. Fortunately Judge Coxe had a fine sense of humour that was tickled by the situation, but he appointed Judge Marsh, genial, portly and silvery haired, to preside over the depositions and act as referee. The appointment of Judge Marsh as Special Master created a precedent frequently followed in subsequent cases.

Thus the remainder of Ferguson's evidence was taken in relative peace, but the gruelling ordeal went on five days a week until early December. The Ford lawyers hoped to wear him down; they asked him about 60,000 questions on every minute aspect of his working life. His answers filled almost eleven thousand pages of verbatim report. He gave a masterly performance despite his irate outbursts. Judge Marsh described him to Judge Coxe as 'the most interesting personality I have ever met in my life'. His own lawyers, despite their difficulties in dealing with him, were impressed by his integrity. Once, when it was thought that Ferguson's telephone at the Gotham Hotel might be tapped, one of the lawyers frivolously commented that a tap on Sands's phone might be interesting. Taking the remark seriously, Ferguson said quietly but firmly, 'I couldn't let you do that.'

After a gay dinner with his lawyers to celebrate the end of his ordeal, Ferguson, who was overtly jaunty but in fact deeply tired—as one would expect a sixty-five-year-old man to be—

sailed for home. Only the fact that he was engaged in a fight kept him buoyant for those three months of examination.

Back in England, he received the congratulations of his lawyers and the news that the Ford attorneys seemed even more worn by the examination than he was. One lawyer, Duggan, had fallen ill.

'It was certainly interesting to hear that Duggan got sick,' Ferguson wrote in reply. 'Of course, you sent him a bunch of flowers with my good wishes! Should Seymour fall sick, send him a wreath and tell him we are living in high hopes!!'

While the long-winded procedures of pre-trial investigation were going on, dazzling events were taking place in the business affairs of Harry Ferguson Incorporated. It has been mentioned that the launching of the case improved morale in the Ferguson team, but it also became something of a *cause célèbre* in the public eye with the result that the re-establishment of the Ferguson business became easier. And Ferguson used the publicity of the case to maximum advantage in rebuilding his company. The first practical step taken in 1947, after getting rid of Kyes and making D'Angelo chief executive, was to arrange for 25,000 Coventry-built tractors to be imported into the United States as a stop-gap pending a new production facility. Quite apart from providing a temporary solution to the North American dilemma, the dollars earned under this plan earned Ferguson the gratitude of the British government and fully vindicated Sir Stafford Cripps's sanctioning of materials for production at Coventry. In January 1948, Harry Ferguson Incorporated purchased a 72-acre site in Detroit on which to erect an assembly plant. The site was bought under the pseudonym of Mr. Harlan because another site, for which negotiations had been in progress a few weeks earlier, had been mysteriously withdrawn from the market at the last moment, an event which the Ferguson management thought related to the litigation with Ford.

Time was the essence of importance in building an assembly plant and getting the first tractors off the line, for the imports from Coventry were only satisfying part of the market and the distributor demand. Excavation on the site was begun on February 13th. Ferguson thought that three months should be enough for the construction, though the contractors protested that they could not complete before the end of July. In fact the plant was finished on July 26th, only 116 days after work began. During the construction period, the Ferguson management busily hunted for component manufacturers and signed agreements with firms like Borg-Warner, who supplied transmissions, rear axles, and hydraulic pump assemblies. The executive in charge of procurement of components was Albert Thornborough. He carried out a difficult task brilliantly; it was difficult because many potential suppliers were afraid that an agreement with Ferguson might jeopardise business with Ford. Thornborough showed remarkable ability—ability that subsequently took him to the presidency of the enormous Massey-Ferguson corporation.

On October 11th, 1948, Harry Ferguson travelled from New York to Detroit to drive the first tractor off the production line at the new plant. It was a memorable occasion: less than a year after the nadir of November 1947, when the company seemed doomed to extinction, it was reborn in the form of an elegant modern synchronous flow assembly plant from which a gleaming little grey tractor had just rolled down. The plant was soon prospering, for by the end of 1948 it was assembling a hundred tractors a day, and a sales network was also being re-established to replace the distributors who had gone over to Dearborn Motors. The future seemed assured.

This progress had a marked effect on the implications of the lawsuit because most of the damages claimed were based on the fact that the whole Ferguson business in North America had been ruined; but now it was gathering momentum again. 'Instead of having a corpse on our hands, we had a convalescent,' Sonnet remarked dryly. However, this did not alter the legal tussle to any extent. Both teams of lawyers continued to burrow their way through mountains of documents; witnesses were examined in minute detail, among them Sands who was put back on the stand in early 1949 for six weeks. The Ford lawyers seemed to be prolonging the proceedings as much as possible in the hope that Ferguson's money or will to fight would run out. It was a vain hope because Ferguson declared that he would fight even if his last penny was spent in the effort to assert the rights of a small company in the face of a giant of industry. Despite this declaration, however, he was constantly worried about the drain on his resources and urged his lawyers to push for more action.

In July 1949 Ford filed a counter-claim against Ferguson, probably in the hope of edging him towards a settlement. The claim held that Ferguson had conspired to dominate the world tractor market and run the Ford Motor Company and that several Ferguson patents were in reality invented by Sands. Ferguson was incensed by the Ford assertions that he had appropriated Sands's inventions for his own uses. 'Now, Mr Seymour,' he said, 'I have been through the mill in inventing, and let me tell you in all friendliness, that any fellow in this world who produces a good invention does so only at the cost in toil, sweat, blood and tears and sacrifice and hard work, and sleepless nights that only that man could describe; and only that man could appreciate what it means when he is told that he really stole that invention from someone else, from some other poor fellow. If indeed I did steal anybody else's invention, then let me tell you I would look upon that, in view of the sacrifices needed to get an invention, as one of the foulest things that a man could do.' The Ferguson lawyers filed a reply stating that the counter-claim was 'wholly without merit'.

By August, feelers for a settlement were being put out and lawyers from both sides had many conferences. Ferguson was not really in favour of any settlement despite being reminded of the old adage that 'a bad settlement is better than a good lawsuit'.

Nevertheless, Henry Ford felt sufficiently encouraged to believe that a meeting with Ferguson might produce a settlement, and he therefore let it be known that he was willing to visit Britain for such a confrontation. When Ferguson heard this he decided that a private meeting between them at his beautiful country home, Abbotswood, near Stow-on-the-Wold, would be the most likely to produce results, and he wrote as follows to Ford:

'I understand it would not be possible for us to discuss any general settlement terms without our lawyers, but we can discuss the world situation and how it can be remedied. This, whilst giving us an opportunity to serve mankind, constitutes the greatest industrial opportunity in the history of commerce. The greater the service to the people, as your grandfather often remarked to me, the greater the industrial opportunity. You know that he believed in the Plan behind our System.

'The world situation today is one where, if statesmen and industrialists do not combine to meet world destitution, Communism will triumph and everything in this world worth having will be lost. I feel sure you will agree that our talks here should cover this background and the vast industrial opportunity presenting itself.

'By studying and discussing the world possibilities for doing good we can best decide whether we could work together, or whether we should go on fighting.'

Henry Ford replied that he was disturbed that no lawyer would be present. He wanted it understood that he was interested in settling the litigation on a mutually satisfactory basis that included 'no merger, partnership or a joint or mutual enterprise'.

'I am deeply appreciative of your invitation to Abbotswood,' he wrote. 'Under the circumstances, however, it seems to me appropriate that our conversations be held in London. After the settlement has been agreed upon, I hope to be able to accept your kind invitation to visit Abbotswood.'

Ferguson was so annoyed that he considered not seeing Ford at all—anywhere: he was unable to put himself in Ford's shoes and realise that an opponent's house and hospitality hardly constitute a position of strength from which to negotiate. For it was obvious that philosophising over the world's evils would in the end have degenerated into talk about more concrete matters. Eventually, he allowed himself to be persuaded to see the 'young man', as Ford was referred to in telegrams between Ferguson and his lawyers, and Ford set out across the Atlantic with a retinue of legal and financial advisers. He made the mistake of telling the press the purpose of his visit, however, and this too annoyed Ferguson at the time. Later he was pleased because he believed that Ford had weakened his position, at least in public eyes, by making a special journey to Britain to see his adversary.

The meeting was to take place at Claridges and the opposing teams booked in there; Ferguson's consisted only of himself, John Sonnet, John Turner and Horace D'Angelo; that of Ford was very much larger. Ferguson laid down a blunt and simple strategy for the meeting: if Ford made an initial offer of less than a certain figure—probably $25–30 million though no one remembers exactly—they were all to get up and leave *immediately*. He was not prepared to start haggling at a low figure and thus prejudice future negotiations or the lawsuit. He himself only planned to speak 'briefly and emphatically on points of high principle, and the details would be left to the others'. This whole approach was typically Ferguson—scoring every bar of the music for full orchestra.

After messages had been carried back and forth on silver trays between two floors of Claridges by red-tunicked and white-stockinged waiters, the first meeting was convened to take place in room 335. Courteous greetings complete, Harry Ferguson perched alertly in the biggest armchair while Sonnet and Seymour got down to some preliminary sparring. The omens seemed good and there was a sensed willingness to compromise between the lawyers, a feeling perhaps stronger than words could have conveyed that the talks were leading towards a conclusion.

It was Henry Ford who became restive first and interjected that the time had come for some concrete details to be settled. And he mentioned $10 million as his opening offer to settle the litigation. Ferguson was immediately on his feet. 'I can't accept, gentlemen,' he announced pleasantly. 'Good-bye, Henry. Give my regards to your mother.' And he walked briskly out, followed by his team. Even they, despite their briefing, were surprised. The Ford delegation was completely dumbfounded. Journalists waiting in the street gathered around Ferguson and he told them jauntily that the Ford offer had been 'totally unacceptable'. 'What are you going to do now?' one of them asked. 'I'm going home,' he said.

The walkout at Claridges was a fine tactical gesture within the overall Ferguson strategy of intractability; but there can be no doubt that Ford really did want a settlement and might have paid over $15 million and a royalty on each tractor produced to have it. Once the attempt to reach agreement had so rapidly proved abortive, there was no further course open but to return to the dreary gathering of pre-trial evidence. Examination of Ford witnesses began. Henry Ford himself, commuting from Detroit, began to give his evidence on October 26th, 1949, and continued until the end of the year. The main line of investigation taken by John Sonnet and his colleagues was in connection with the apportionment of Dearborn Motors stock among Ford and Dearborn executives. A piece of paper with the names of Ford executives and a figure after each, written in Henry Ford's handwriting, was the cornerpiece on which the Ferguson lawyers constructed their evidence. Ford was frank in his admission that the figures did represent stock allocations and that he had probably written them before the break with Ferguson. He also admitted that, even if Fords had for some time made a loss on tractor production owing to wartime price controls, the operation was profitable at the time of the break with Ferguson.

Another fundamental issue which was exhaustively examined was a report by a Ford employee addressed to his management in 1946. It was a study of tractor, implement, and parts distribution, outlining the capital cost and likely return, and it was made, as the employee stated, 'with the thought in mind that we would continue the distribution programme as it presently functions under the Ferguson organisation using the same distributor and dealer set-up'. The report admitted that 'Ferguson might want to stay in business and tie up implement manufacture to make things difficult' but that this should create no real problem.

The gathering of evidence went forward with excruciating slowness. Ferguson's lawyers still believed that Fords were employing delaying tactics. On one occasion a Ferguson lawyer jumped up during a hearing and said, 'I'd just like one last word on that subject.'

'Come around again in 1960,' a Ford lawyer retorted cheerfully.

This was perhaps nothing more than ready wit and an awareness of the snail's-pace progress being made, but Ferguson's lawyers took the remark as being indicative of Ford's procrastination.

Then in March 1950, Philip Page, a Vice-President of the Ferguson company, took the stand after four days of briefing. He was thirty-six years old and had received his training with Ford before joining the Ferguson organisation in 1944. He was obviously under stress during the examinations and he was very anxious to do well. After being involved in a taxi accident, in which he received a blow on the head, and passing some sleepless nights too, he jumped from the window of his fourteenth-floor room at the Hotel Pierre. The note he left for his wife betrayed all the agony of a loyal man whose allegiances were tearing him asunder: 'My head is tight as a drum. First time I recall. I don't know why. Tried hard, honey. Best I know how. Can't crucify Ford, they gave me my training. Another compromise but know we have the best product—too bad they couldn't reconcile their differences. Now I leave you and the kid with my life. Have broad shoulders—broader than mine.'

The tragedy of Page's suicide appalled everyone concerned with the case. Some were moved to make an inward appraisal of the real issues at stake, of the validity of such a power game if it drove a man to suicide. But if they had qualms, they hid them well and proceedings resumed.

In the summer of 1950 the whole legal circus moved to Britain to gather more depositions, at Ford's request. Examinations were conducted in Belfast, Leamington Spa and London. Ford had to assume the cost of this expensive exercise which involved Judge Marsh, thirteen lawyers and a team of court stenographers. A battle of documents took place in Belfast when truck-loads of papers were being trundled from one team of lawyers to the other in their different hotels. One canny Ulsterman quickly invested in a photocopier and paid for it in a few hours as each team tried to swamp the other with paper.

In London, Harry Ferguson was again called to give evidence for about two weeks. During that period, Seymour wrote a note to Ferguson:

> Mr. Ferguson:
> This is to confess that we all *love* you. My examination was what we will call in the 'line of duty'.
>
> Whitney North Seymour

For this truly extraordinary letter no explanation has been available, unless it was meant to remove some rancour of that day in the examination room and pave the way for a more amicable exchange in the future.

Finally, in the autumn of 1950, most of the evidence had been gathered and plans for the trial itself were being laid. Ferguson wanted the jury to see a demonstration of the Ferguson System and even had his lawyers find a suitable place, a National Guard armoury (at 94th Street and Madison Avenue) that was lit, heated, had a capacity of 1500 people and an earth floor he could plough up. Ford lawyers selected no less than 6500 documents to bring to court as exhibits; Ferguson's team wanted to present 1200 as evidence. Believing that at long last the case was really coming to trial, Ferguson was less inclined than ever to negotiate a settlement, despite several round-about approaches from different quarters, and urgings from his lawyers. In mid-December he went to New York to see to business matters and to goad his team into a suitable frame of mind for battle; for he considered that all this talk of settlement denoted a weakening in their resolve to fight. While in New York, he did allow his lawyers to write to their Ford counterparts to say that he would be willing to consider a settlement for about $60 million, a figure which John Sonnet and his colleagues held to be unrealistic. Since its establishment, Dearborn Motors had made about $23 million profit and Sonnet believed that a settlement for about $25 million would be reasonable.

The date of the trial also presented problems. To hold the proceedings before a jury would have caused a delay of up to six months. There was an underlying feeling too that it would be unfair to subject a jury to such a lengthy and complex hearing. Ferguson was by now exasperated over the delays; the legal costs were already astronomical; each passing month saw more Ford tractors, incorporating his patents, produced and sold; and those patents themselves were also marching steadily towards their expiry. Finally it was decided to waive the right to a jury trial and the case was set to begin in March 1951. By then, Ferguson had increased the sum of $250 million mentioned in the complaint by a further $90 million to cover patent infringement on the Ford tractors built since the complaint was first lodged, so about $340 million was the sum being claimed—unrealistically in view of the manner in which the Ferguson company was prospering in North America.

In his opening statement in the court, Sonnet laid great stress on the large amounts of money made by Ford executives through Dearborn Motors. He quoted Breech as having invested $25,000 in Dearborn shares at the end of 1946; by three years later he had received $250,000 cash in dividends, and his shares were worth

$2·25 million—a 10,000 per cent return on his investment. The Ford lawyer, Bromley, also made an effective opening statement, but thereafter the case dropped into mind-boggling detail of the sort that had characterised the pre-trial investigations.

After the cross-examination of Ed Malvitz, who it will be remembered had helped to lay on the demonstration that led to the handshake agreement, Horace D'Angelo took the stand in May and began a court-room marathon that is thought still to be a record. Judge Noonan, who was presiding, was frequently in despair at the lack of progress. 'Are you proceeding on the theory that old witnesses never die?' he asked the Ford lawyers. 'This witness is growing old before my eyes.'

Finally, with D'Angelo still on the stand at the end of June, he decided to adjourn from July to September. He remarked that a senior judge had said that if he did not take a holiday he would make such friends with both Ford and Ferguson lawyers that they would contribute to the cost of erecting a tombstone to him at the end of the case. He also suggested, in jest, that since a trial before a jury would certainly be shorter, they should consider the present proceedings a mis-trial and begin again before a jury in October. Judge Noonan was to a certain extent disenchanted with both sides for their intractability over a settlement. He was also unfavourably impressed by the character of Harry Ferguson when he learnt that in 1946 he had been soliciting secret reports on Kyes's performance from other directors.

Then, with the court adjourned in July 1951, Harry Ferguson suddenly succumbed to the urging of his lawyers and agreed that they should try to settle the case. In a letter dated July 17th, he wrote to his lawyers.

'Gentlemen: TOP SECRET

MY PRESENT FEELINGS

I think that before setting down some details of the course which we should follow I should express my feelings on some matters bearing strongly upon our lawsuit.

Rightly or wrongly, and for our guidance in the future, I feel we made a mistake in giving up the jury. It would appear to me, rightly or wrongly, that the abandonment of the jury opened the flood gates to the Ford lawyers to extend this trial far beyond the period of time it would have taken had a jury been retained.

This, in turn, makes the cost of the litigation so high, and what is worse, it enables the Ford Company—or helps to enable them—to extend the period of trial for so long that all the patents in the suit of substantial value to us would be expired and so the great cause for which we have been fighting really no longer exists.

In other words, I am bitterly disappointed and discouraged by the fact that, because of the unfairness of the law towards inventors and inventions in your great country, those who break the law have great advantages in their favour in a lawsuit such as ours.'

What Ferguson was fighting for was the rights of the inventor to what he had created in toil and sacrifice. 'We could have won that fight if we could have carried our lawsuit to a point where we could have stopped the Ford Company from infringing one or more of our patents, but as we cannot now do that the splendid and fundamental ideals behind our fight have gone and I, therefore, after long and full consideration with Mr Sonnet and Mr D'Angelo, have come to the conclusion that if a reasonable settlement could be made with the Ford Company, we ought to make it.'

Harry Ferguson's new attitude was such a *volte face* that it is interesting to speculate on what caused it. Some authorities, including Professor Neufeld in his book about Massey-Ferguson, *A Global Corporation*, have suggested that the involvement of a top Ferguson executive in a divorce case so tarnished the shining morality of the Ferguson Company that he decided to seek a settlement rather than continue the fight.

The extract quoted from Ferguson's letter to his lawyers, on the other hand, betrays a man deeply discouraged. It seems more likely therefore that black despair overtook him. The reality that the case could go on for years, and the patents expire in the meantime, forced itself upon him. The whole fight was pointless and, if the past lack of real progress was any indicator, probably impossible to win. In addition, Sonnet was urging him to settle because secretly he believed that Ferguson could not stand the strain of appearing in court. He thought that the Ford lawyers would probably needle him—by casting doubts on his ability and on his role in inventing the Ferguson System—to the point where his anger would create such a scene in the court that his case would be irretrievably lost. And perhaps he would also have suffered grievous harm to his health and well-being in the process.

Once the decision to settle had been reached it was only a matter of patient negotiating. This went on behind the scenes while D'Angelo continued to give his evidence when the trial resumed in October. He was not released until he had been under examination for over forty days in total; and then John Chambers went onto the stand to testify. The out-of-court negotiations almost broke down completely on several occasions, notably when Fords offered $1·5 million compared to the $15 million demanded by the Ferguson side. Harry Ferguson considered the offer 'despicably mean'. Finally, after the most elaborate proposals and counter-proposals, Judge Noonan selected a figure of $9·25 million representing patent royalties and agreement was reached on this consent judgement. The agreement included the withdrawal of all claims, including Fords' counter-claim, and the concurrence that as of the end of 1952 Ford would cease to manufacture tractors incorporating some of the Ferguson patents, the most important being hydraulic control on the suction side of the pump.

Thus a litigation which had involved over 80,000 depositions, over a million documents, 200 lawyers, and cost Ferguson about $3·5 million came to a close. Ferguson was embittered by what he considered a travesty of justice and a betrayal of the rights of the little man. He had enjoyed the fight for most of its duration; a journalist once asked him whether he would not be glad when it was over. 'No, why?' he replied. 'Some people play golf, some fish, I have my case.'

When it was finally settled in April 1952, however, he arranged that the consent judgement award should be represented as a great victory. He was never known to miss an opportunity for good publicity, and he carefully concealed his disappointment as he stoutly declared that the award upheld the rights of the small inventor in the face of giants of industry. The press widely carried the story of the 'victory' that in his heart he felt to be a defeat.

In one sense, however, it was no defeat: the case had provided clamorous publicity and this, coupled with the excellence of its products, carried the Ferguson Company into a pre-eminent position. In the Eastern Hemisphere, the Coventry-built tractors were selling in larger numbers than those of all other companies put together. During parts of 1949, for example, Ferguson had no less than 78·4 per cent of the wheeled tractor market of Great Britain. The lawsuit, with its inherent appeal to the sympathies for the underdog, must have been of inestimable value.

PART THREE

25 *Country Gentleman and Business Tycoon*

Some great executives in business and political life have shown that a well-rounded approach to living is the most productive. Probably the supreme example is that of Sir Winston Churchill who partly through his painting and bricklaying found the relaxation necessary to enable him to shoulder the heavy responsibilities of his work. The fact that he also wrote more in his lifetime than most professional writers is an indication of the manner in which he spread his remarkable energies over different fields and yet excelled in several of them. Henry Ford shrugged off the stresses of his working life by meticulously dismantling and repairing watches in the calm of his home, as well as by camping trips, and by his interests in square dancing, and in American folklore and history generally. Onassis sails, swims, water skis and leads a busy social life all over the world. To work without relaxation, without distractions to replenish energy and give a fresh view of matters seems to be counterproductive in the long term. That dedication is instrumental to success is obvious, but dedication should be applied to more than one element in life. Yet Ferguson had never developed any real interests beyond the work to which he gave every moment, including those many added by insomnia.

Then, finally, when he returned to Britain after the war with his wife and daughter he found the outlet he had always lacked. Having spent their life in hotels and rented houses, frequently moving to satisfy the exigencies of business interests, they decided to buy a country home within comfortable reach of Coventry, where Harry Ferguson Ltd. had established its offices and engineering department on the Fletchampstead Highway. A beautiful Cotswold house and estate were found at Stow-on-the-Wold; and this home, Abbotswood, became a sheet anchor to Ferguson in the stormy struggles of his post-war business activities and when increasing depressions afflicted the latter years of his life. The fine stone house, owing much of its elegance to some remodelling by Lutyens in 1903, lies in a fold of the hills and is surrounded by parkland sweeping down to a crystal trout stream, the Dikler. Exquisite gardens and lawns encircle the house with its mullioned and creeper-fringed windows.

Abbotswood was the first real home of the Fergusons' married life, and they settled into its beauty with a deep feeling of content. They brought Irish staff to run the house and to give it a close link with Ulster, about which Ferguson was becoming increasingly sentimental with advancing age. He was even nostalgic about the bigotry of his native land and when a child of one of the staff chalked 'To hell with the Pope' on a farm gate he was highly pleased and remarked that it made him feel really at home. The inculcated prejudices of his own childhood had travelled undamaged with him over the busy years, even if he had the slogan removed at once from the gate.

Abbotswood was also an outlet for his perfectionism, which he carried to cranky extremes. The estate staff had to clean the Dikler of any water weed growing in it and of any rushes along the bank. At one point along the stream he wanted to excavate an artificial lake that would shine silvery in the green of the parkland. But it had to be perfectly circular when seen from his bedroom and there had to be an island in the middle which divided the flow of the Dikler into exactly matching swirls of current. The excavation continued for weeks with regular visits to Ferguson's bedroom by the estate staff in order to be able to adjust and replan the lake's contour.

There was no question of nature being allowed the right of self-determination at Abbotswood. The trees had to grow bolt upright to please their owner, and if they did not they were guyed into line. The lawns were not to be besmirched by a single dead leaf, and in autumn it was almost a full-time job for one of the estate staff to sweep them up as they fell. A neighbouring farmer was allowed to graze his sheep on the parkland below the house, but only if they were not shedding tufts of wool; and even then Ferguson was annoyed by the way they wore tracks across the grass as they followed in each others' footsteps. He therefore had steel hurdles made up with a spike to be driven into the ground to hold them; when a path began to form his staff had to block it with hurdles in order to divert the sheep.

Ferguson's pastime when nothing urgent called was that of either mowing or rolling the parkland. A tractor was kept ready for him at all times and when he wanted to relax, think and plan he drove systematically around the parkland for hours on end. He was very particular that rolling should be done when the ground was at its wettest, and he therefore went out frequently while it was raining and drove the tractor and roller to and fro. An array of waterproof clothing completed the equipment for such occasions because he was convinced that getting wet was the only cause of the common cold. Infection from others was not an admitted possibility, and hence any employee suffering from a cold was guilty of gross carelessness. Even influenza and other minor ailments he regarded with the suspicion that they might have been induced by negligence on the part of the sufferer. John Peacock, an accountant who faithfully served for many years at Abbotswood, the nerve centre of the group of Ferguson companies established after the war, once fell ill when his employer was in the United States; with tongue in cheek he referred in a letter to his 'sin of sickness'. If Ferguson thought that this gentle tilt at his views was funny, he did not bother to say so. On the other hand, Ferguson was capable of great thoughtfulness. Again while in the United States, for the lawsuit, he heard that the managing director of the Coventry-based company, Alan Bottwood, was suffering from rheumatic pains of the shoulder. He at once wrote to the butler at Abbotswood asking him to send his sheepskin coat to Bottwood in the hope that its warmth would relieve the pain.

There was a Jekyll and Hyde quality about his relations with his staff. His punctiliousness at Abbotswood was almost beyond belief. His shoes had to have their laces put in with an exactly equal length of end remaining; his bed had to have an exactly equal overhang of sheet on both sides; meals had to be exactly on time and perfectly served. If all the harsh standards were not precisely adhered to, a pained reprimand resulted. Yet at the same time he insisted that the domestic staff be let off early in the evening and that not too many house guests in succession should tire them. He once explained that the domestic staff were not servants but part of his team for making a better world; his butler's services, for example, formed a key part, as did everybody else's.

In Abbotswood there is a fine panelled library with a bow window looking out over the formal gardens and thence over the sweeping parkland. This dignified room with its comfortable sofa and chairs, large fireplace, and desk set in the window, was Ferguson's Abbotswood office. As Ferguson companies were established in other countries such as France, India, South Africa and Australia, the hub of the empire concentrated increasingly at Abbotswood, and Ferguson kept a nucleus of staff there to deal with the centralised financial matters of the holding company.

His own secretary, who was always male, put out volumes of Ferguson correspondence and directives. These letters were always laid out with headings typed in capitals because Ferguson insisted that new subject matter should be properly introduced in the interests of immediate clarity. He was utterly incapable of writing a short letter, and epistles of incredible length and on every topic imaginable emerged almost daily from Abbotswood. One series of letters addressed to a variety of companies was provoked by the fact that the jets of the twin fountains in the garden were uneven; another that continued for years was to a shop in Belfast that sold coffee. 'COFFEE, COFFEE, COFFEE!' was the heading for one letter in which he went on to say: 'However, I am most deeply grateful to you, and so is my family, for all the trouble you have taken to help us in this vital matter of coffee which, believe me, makes all my big business ventures and such things as the Ford law-suit fade into insignificance.' (Not to be outdone in blatant blarney, the shop replied, 'Sure the honour of attending to the wants of an illustrious Belfastman who is now an international figure is thanks enough.')

Life at Abbotswood was governed less by the clock than by the second-hand of the clock. This is of course an exaggeration—but it is no exaggeration that Ferguson called one of his staff on the internal telephone and asked him to come to the library not at 9.30 or 9.35, but at 9.32. And woe betide anyone who arrived late at Abbotswood for an appointment; one steel company magnate drove up to the front door a little more than five minutes late after a journey of about two hours and received the message that, since he was late, Ferguson could not see him. News of this idiosyncrasy travelled fast, and it was a common sight to see cars parked on the verge of the road leading from

Stow-on-the-Wold to the estate of the 'terrible pernickety man'; their occupants were awaiting the precise appointed time to glide down the asphalt of the drive flanked by its perfect lawns and majestic trees and swing to a halt before the portals of the imposing house. If the guest was of some importance, the small grey figure of Harry Ferguson, with only its erectness to match the grandeur of the surroundings, would be waiting on the doorstep to extend a courteous greeting.

When Ferguson was being driven through the countryside in his Rolls-Royce, he usually sat next to the chauffeur, unless he felt like a catnap, in which case he would hop over the seat and stretch out in the back. But if he was sitting up, he often scanned the fields along the roadside for Ferguson tractors at work. If he saw one of his brain children that was not performing properly he would tell the chauffeur to stop the car. After a critical stare over the gate to confirm that all was not well, he would walk towards the tractor and imperiously hold up his hand for the driver to halt. Sometimes without even bothering to identify himself, he then delivered a lecture on the use and maintenance of Ferguson equipment, making adjustments with a spanner as he did so. And if the tractor was dirty and uncared for, he had no compunction about telling its operator to mend his ways. There is no record of any tractor driver ever telling him to mind his own business, for even if he was small in stature, his demeanour was authoritative.

About twice a week Ferguson was driven to Coventry. Precisely a mile from the offices and engineering department of Harry Ferguson Ltd. on the Fletchampstead Highway the Rolls drew in to the kerb and Ferguson got out. He believed that everyone should walk at least a mile a day and he briskly practised what he preached. On reaching the headquarters his first steps usually took him straight through the main part of the building to the engineering workshops behind. There he would ask at once about progress on new implement designs and wander around to look for himself. Coming up beside a draughtsman at his board, he would put his hand on the man's arm to move him aside, saying, 'Let me see, my boy.' Blue eyes behind gleaming lenses passed in critical scrutiny over the work, while the draughtsman stood mutely expectant of the best or worst.

Ferguson's employees had to follow his own habit of always having a notebook and pencil ready to jot down sudden inspirations or tasks to be done, and he was likely to stop an employee in the corridors of the company offices and ask to see his notebook. Not only did the employee have to produce it for inspection, but it had to come from the correct pocket; for in the interests of efficiency it was to be carried in the left jacket pocket with the pencil in a pocket where the right hand could reach it easily. Thus in one fluid motion of both hands, notebook and pencil were at the ready.

Apart from the engineering activities at Coventry, Ferguson's main interest lay in the publicity department headed by Noel Newsome. Newsome had been a journalist with the *Daily Telegraph* until World War II, when he was summoned to the B.B.C. and told that he was in charge of European broadcasts. From a skeleton service that had been broadcasting two programmes daily in German, French and Italian, Newsome was responsible for building up the network of news programmes that reached into every corner of embattled Europe in a multitude of languages, the dramatic drumbeats heralding the messages that brought hope of relief to millions of oppressed people and resistance fighters.

Like all new executives joining the Ferguson Company, Newsome was first sent on a course to learn about the technicalities of the product. Training was, as usual, of prime importance in Ferguson's plans and, as soon as the Coventry-based company was established, courses began in the grounds of Packington Hall a few miles away. Subsequently the Company leased part of Stoneleigh Abbey and its parkland and in these fine surroundings set up an exceptionally sound training school to which students flocked from all over the world.

The courses given at Stoneleigh were characterised by an astonishing candour about the Ferguson equipment, and by a fervent attention to accuracy in the giving of information. In fact it was easy for the instructors to admit freely to the very few weak points in the equipment: its strong points were so outstanding and numerous. Ferguson tractors and implements were the success story of post-war mechanisation on British farms and on farms throughout much of the world. Apart from one or two difficult periods of export sales, caused mainly by such extraneous matters as exchange control and import/export regulations, Ferguson equipment was selling as fast as it could be produced. There was therefore no need for the instructors at the Stoneleigh School to worry about a sales bias to their courses; it was sufficient to be meticulously accurate and well organised, and the sheer excellence of the product, which at the time had no rivals, sold itself. The enthusiasm and dedication among the instructional staff and their groups of trainees was amazing. Thus, quite apart from the wealth of knowledge and practical experience of the Ferguson System gained on the courses, they were also a fine introduction to the crusading spirit of the Company and to the emphasis on attention to detail, accuracy and integrity. There was a remarkable sense of purpose in the Company, not the purpose of making money, but the purpose of making a first rate product and of helping the world to solve some of its problems. The fervour, as noted by many, was all-pervading and irresistible.

Noel Newsome had not yet met Ferguson when he began his course of indoctrination at Packington—he joined before the establishment of Stoneleigh—but after a couple of days he received the message that the 'old man' wanted to see him. He took off his blue trainees' overalls and in corduroys and leggings made hastily for the Fletchampstead Highway offices. Ferguson was in the boardroom. He took a frosty look at Newsome as he entered. 'Of course,' he said, 'one of the things you have got to realise working for us is that you must dress properly.'

'I've just been called from the field. I don't normally dress like this,' Newsome replied.

'All right, but you're too fat. You ought to do something about it.'

Little else was said at that rather intimidating first interview, but subsequently Newsome became one of Ferguson's most ardent admirers and an expounder of Ferguson doctrines on economics. He handled the public relations work of the company with flair and dignity, two qualities often difficult to reconcile in the world of publicity. Ferguson thought highly indeed of Newsome, whom he called upon to handle a variety of problems, but there was almost a parting of the ways once over a banal incident concerning type faces. It had been decreed that not more than two different type faces were to appear in any one Ferguson advertisement, the purpose being to strike that note of functional simplicity that was also the hallmark of the company's engineering designs. Owing to a misunderstanding, however, Newsome ran one advertisement with the prescribed two type faces, but used several different sizes of each. When Ferguson saw it he was so angry that he did not speak to Newsome for several months, ignoring his very existence, apart from remarking to Alan Bottwood, the managing director, that he had no further use for Newsome's services.

The storm blew over ultimately and Newsome once again became one of Ferguson's close associates in the high pressure but sophisticated public relations activities of the Company. He found it stimulating to work with a man who was himself such an able promoter. Indeed, Ferguson had once pulled off what one newspaper termed 'the greatest publicity stunt of the forties'. The event occurred when, in 1948, Ferguson had decided to give a large cocktail party in order to promote the export sales of his equipment. He decided that the ballroom of Claridges would be the most suitable place for such a gathering, at which a tractor would be on show. But the management were horrified at the idea; it seemed almost a desecration of their dignified institution. Only an appeal to their sense of patriotism, explaining the benefit that such a function could have on Britain's beleaguered gold and dollar reserves, finally overcame their reluctance. Grudgingly they allowed that the tractor might be brought in through a side entrance, which meant partially dismantling it and reassembling it in the ballroom.

During the reception the tractor, complete with plough, sat on a dais about as incongruous amid the splendour of brocade curtains, shimmering chandeliers and gleaming floors as a sheep dog that had strolled in on a state banquet. Ferguson made a speech extolling, as usual, the virtues of his System and what it could do for mankind. A Russian expressed scepticism over some of the claims of manœuvrability for very small plots of land; this was precisely the opening Ferguson needed. He climbed onto the tractor, started it and deftly executed a number of dainty pirouettes where, on less cataclysmic occasions at Claridges, bejewelled and bemedalled royalty and nobility elegantly swirled. Having swung the landside of the plough within a millimetre of the brocade curtains he drove the tractor into the lobby, calling for the startled people to clear the way, and bumped down the steps into the street. Few other stunts could have equalled this for attracting the attention of the press. *The Guardian* commented in an editorial that only 'presenting the thing at Court' could have had a greater impact.

One may easily wonder how a man like Ferguson inspired so much devotion and service from his staff. Few employers can have been more demanding and intransigent, and he was parsimonious with praise too. The secret lay in the cause that was being worked for, and in his great compassion for humanity. The more closely one worked with him the more evident the latter became. Beneath that austere and sometimes bleak exterior, showing only occasional flashes of whimsical humour, lay a core of concern for others, a desire to see less misery and poverty afflict mankind. He was not over fond of animals, yet one day, before the Phoenix, Arizona, demonstration in 1942, he arrived in the field and found some of his men encircled around a pair of gophers that they were trying to incite to battle. 'Leave them alone,' said Ferguson angrily. 'There is enough suffering in the world already without adding to it.'

Poverty and the human predicament preoccupied him even at a distance. He was a radical in much of his thinking, a progressive on behalf of the toiler on the land. But he was also a violent anti-Communist at the same time. Communism was inevitably Stalinism by his lights, and it was the great evil of the world in its suffocation of individual enterprise and liberty.

There were certain endearing traits in his behaviour with employees which helped to counteract the fussiness, the demands, and the not infrequent bouts of apparent ingratitude that drove loyal staff to the limits of silent exasperation. For example, when someone went to see him he would often pull a chair up beside his own and pat its arm inviting the employee to be seated and enter into the intimacy of what he would term a 'nice chat'. And even when he took it upon himself to explain to someone how to tie his tie or fold his handkerchief, the recipient of the lecture usually found it impossible to be irritated; for Ferguson always managed to give the impression that he was taking pains for the benefit of the listener, that he was doing him a favour.

Perhaps most significant of all, however, was his infectious enthusiasm and total self-assurance. In very many ways he was not unlike Montgomery; both forceful men, they also shared an ability to cast a spell through quiet insistence on setting great objectives and through their confidence that success would be achieved. The cause was good; the crusade might be long and hard; but victory was essential and never seriously in doubt. And, as with Montgomery, the discipline imposed by the leader, and his own extraordinary attributes and idiosyncrasies, gave rise to a fund of stories about him, stories that were often embellished and exaggerated, but all of which dwelt upon his overwhelming self-confidence, his fussiness and the patriarchal manner in which he oversaw the activities of his staff, always convinced of his own indisputable rightness. Perversely, these stories increased the liking and admiration for the man, partly because normal people who have not the self-assurance to step beyond the bounds of conventional behaviour love a successful eccentric, and partly because courage of conviction is always impressive in the long run, even if the convictions themselves are sometimes open to question.

Perhaps the most important element of all, however, in inspiring the loyalty he did was that Ferguson succeeded in setting up emotional and inspirational currents between himself and his staff. This emerges clearly from a remark passed by Noel Newsome, an intelligent man, and one of a culture superior to that of his employer. 'From the time I met him to the time he died, I was never more under the influence of one man. He had the ability of making you feel that what he was trying to do was the only thing in the world worth doing.'

26 On the Verge of Politics

The economic ideas presented by Ferguson at the Bethesda demonstrations in 1943 were still at that time in their crudest and simplest form. Over the following years he elaborated them until he had evolved a philosophy over pricing and inflation that was much more generic than the mere question of reducing food prices through farm mechanisation—the linchpin of his Bethesda argument. He came to believe that inflation was at the root of all the discontent and unrest in the world; it was a 'seedbed for gangsterism'. There could be no peace in the world, no consistent prosperity until the vicious circle of rising wages followed by rising prices, then again by rising wages, was mastered. The fact that for centuries inflation has been a financial erosion as implacable as that of the soil did not deter him from mounting a campaign to put a stop to it. The need for cheap food produced by mechanised agriculture remained at the heart of his campaign, but many other considerations were introduced.

His attempts to have his Price Reducing Scheme adopted on a national and even international scale involved him in what the *Sunday Times* described as 'a provocative individual intervention in public affairs reminiscent of some of the activities of Henry Ford. Acting independently of political parties, he is campaigning for wage and profit stabilisation as an initial step to force down prices and raise living standards.'

There was good reason to feel concern over Britain's economy. A regular 5 per cent per annum inflation in the four years beginning 1948 raised wages and incomes by about £2000 million without bringing a commensurate improvement in the standard of living. On the international level, his 'plan for prosperity, security and peace' hinged on the abolition of Communism through the abolition of want and poverty. 'If we fight Communism in a war . . . shall we be any better off? The world will be more destitute than it is today. Communism will just be born again, or something worse than Communism. The more we impoverish the world, the more we leave it open to any "ism" that comes along. Those are the conditions under which agitators who promise the impossible get their blind, fanatic following. Our plan ought to be to fight Communism and beat it, not by the usual implements of war, but by beating our swords into ploughshares in the literal sense. We must grow food for the multitudes to eat . . . that is the only final solution.'

But the economic problems most readily evident and calling for attention were those affecting post-war Britain: austerity lingered on year after year, dollar aid was still necessary, and the defeated Germans were marching towards a victory of the peace. 'Prices down—or Britain down!' declared Ferguson and began to bombard politicians with letters about his Price Reducing Plan. One of his earliest interventions was in attempting to persuade Bevin not to nationalise the British steel industry on the grounds that nationalisation would put up prices. His own philosophy over industry and capital was: 'Give the workers the joy of a personal share, financial and otherwise, in the industry in which they work. Give them the pride of working for a reduction in the price of the product which they make—a reduction from which they and everybody else will benefit! If I were fighting an election campaign myself I would present this idea to the electorate, knowing that the workers would prefer to have a personal, financial interest in a successful industry bringing down its prices, rather than an impersonal interest in an industry increasing its prices and heading the nation for disaster.'

He was not, of course, successful in persuading the Labour government to give up its policy of nationalisation, but even so he continued to write long letters to all ministers urging them to endorse his Price Reducing Scheme and make Britain the centre of a new world economy based on the mass production of cheap farm machinery. To Clement Attlee in 1949 he wrote one of his epistles that began with the title line—'THE WAY OUT OF OUR DISASTROUS ECONOMIC POSITION'.

In part, the letter ran as follows:

'. . . Arrangements are being made for me to have discussions with some of the top men in Washington. I intend to put my Plan before them for the saving of Britain. I am led to believe that a plan of the same nature will be adopted in the United States and recommend strongly for Britain. The late President Roosevelt strongly endorsed this plan for a Price Reducing System and would have sponsored its adoption but for his untimely death.

'If, before I go to the United States, you would care to come here to the peace and complete privacy of Abbotswood to discuss my Plan with me I should be delighted to see you. I will convince you that we can quickly set the country on the road to success and security and win back the confidence of the world in doing so.

'Our one great hope is to *bring down the cost of living*. We can only do that through Agriculture. We do not need to lengthen hours nor reduce wages, but we do need to stabilise wages and start the great campaign in Agriculture and Industry for which I have been crusading.'

Despite Ferguson's insistence, the Labour government was not moved to take his proposals seriously. Since he was only interested in objectives, and was unhampered by political allegiances, he decided to attack on the Conservative and Liberal fronts as well. In October 1949 he sent a seven-page letter to every Conservative and Liberal peer and member of parliament. A good deal of favourable comment resulted, though had the letter been shorter it would perhaps have been read more fully. Noel Newsome was responsible for drafting all such letters, but even had he wanted to keep them very short, the elaboration insisted upon by Ferguson invariably increased their length. The increase in length was not matched by an increase in weight and effect however; indeed the opposite was true, because once over two or three pages long the effectiveness of any letter usually diminishes in direct proportion to its length. People either lose the patience to read it through or find its numerous pages so daunting that they never begin to read it, putting it aside for some more leisured moment which never arrives. Ferguson could never conceive that people might not read his letters. It seems to have been his implicit view that their authorship should guarantee them attention.

Just before Christmas 1949 Newsome was at a party in Kent at which Winston Churchill was present. Newsome was related, through marriage, to Clementine Churchill and because of this and his wartime work it was easy for him to buttonhole her husband.

'You know, I think you should meet Harry Ferguson,' he said, 'you'd find it interesting.'

'What, that charlatan who's toting his tractors around?' Churchill rumbled. But despite this inauspicious start, Newsome finally persuaded him to invite Ferguson to Chartwell. The lunch took place early in 1950. Ferguson was accompanied by Newsome, and Churchill asked his son-in-law, Christopher Soames, to attend. Soames was later to become Minister of Agriculture. Ferguson brought his spring-driven tractor model and plough in order to explain the elements of his System. At the luncheon table Ferguson turned the conversation to the fact that the Ferguson Plan and the Price Reducing Scheme needed the leadership of a statesman of world standing to launch it. Roosevelt had given his word that he would put the Plan into operation once the war was over, but he had died too soon to be able to do so; Truman had never been so committed. Now, Churchill heard, the leadership required his own greatness. And he could make it his political platform in the next general election.

'Well,' said Churchill, 'I should be pleased to serve my country again. But don't think I would jump at office like a dog at a bone.'

'I believe you, Mr Churchill, I believe in your sincerity. If I didn't I wouldn't give you ten minutes of my time,' replied Ferguson.

Amusement lit up Churchill's eyes, and he winked the left one (which was hidden from Ferguson) puckishly at Newsome.

'Ah, I see you're a very exacting man,' Churchill rolled out in his majestic, lisping voice. 'I, as you will appreciate, have to be more lenient!' The witticism escaped Ferguson who went on to explain his ideas of defeating Communism through economic betterment of the world. He also suggested to Churchill that he should change the name of the Conservative Party.

'Now then,' said Churchill after lunch, 'I've been looking at your models and things: I'd like you to look at some of mine,' and he lumbered off towards the basement with the little group in tow. Prize among his collection was the original design for Mulberry Harbour, the pre-fabricated concrete breakwater and wharfs that had been towed across the Channel in sections and assembled to provide a port after the allied invasion of Normandy. The design was hanging on the wall of the Chartwell basement. Churchill pointed out the comment written on it by some top-ranking military authorities: 'totally impracticable' had been their initial verdict. After examining this and other Churchill trophies, the group emerged from the basement and Ferguson went to get his coat.

'How did I do?' Churchill asked Newsome in a conspiratorial whisper. Newsome laughed. 'He's a remarkable fellow, remarkable,' Churchill went on. 'Bit of a one track mind though, but remarkable.'

However, it was not only this remark that indicated that Churchill was prepared to take Ferguson seriously. Over the ensuing years, and when the Conservatives had come to power, Ferguson carried on reciprocal correspondence with Churchill and many of his ministers. Indeed, Churchill had hardly become Prime Minister again in 1951 when he received a congratulatory letter from Ferguson. In the last paragraph, Ferguson wrote: 'All we need is leadership. I am sure that if you take up this Crusade of price reducing with the same vigour and brilliance with which you saved the country in the war, you can end your political career, when you want, in a blaze of glory as great as that which crowned your victorious military strategy five years ago.'

Some time later Harry Ferguson Ltd. took a full page of the *Daily Express*. Under the headline 'Prices Must Come Down or We Face Mass Unemployment', the Ferguson plan for Price Reducing and its benefits were explained. The announcement aroused some degree of interest and Ferguson followed it up with another plea to Churchill. 'We can do the job,' he wrote. 'All we need is leadership such as you gave us during the war. Call upon the people and they will rise to that call.'

'Dear Mr Ferguson,' Churchill wrote in reply. 'It would be a great help if you would explain how you would propose that this campaign should be conducted. Leadership is all very well, but the directions must be given with some precision.'

Ferguson answered to the effect that he would need another meeting with Churchill in order to explain the details at length, to which Churchill replied, 'I am very, very busy. Can you not write me the pith of the matter in complete secrecy?' It took Ferguson about a month to write back and say that he could not comply with the request. In fact, the tactics of the price reducing strategy were something that he had never bothered to work out in detail. Just as with his tractor designs, he laid down the overall concept and expected others to provide the detailed mechanism for its implementation. On one occasion, the Duke of Richmond and Gordon, a personal friend, asked Ferguson how exactly he proposed that the Price Reducing scheme should be started. 'Lord, boy,' said Ferguson, 'there are thousands of people sitting on their backsides in Whitehall. They can work out the details.'

By 1952, however, he had laid down certain steps that he believed could be the foundation of Price Reducing. Firstly, all wages and salaries should be stabilised at their existing level; secondly, all distributed profits should be stabilised at the level of the previous full year; and thirdly, all additional company earnings should be used primarily to reduce prices of the company's products, but where they did accrue they should be invested in plant and other inputs that would raise productivity. 'Where earnings are directly related to output, operatives should be allowed to earn as much as they can,' he added, 'and there should be no check to increased earnings due to promotion.'

When it was put to him that such a plan could only be realised by severe legislation he disagreed. A great leader, calling on his people for restraint and understanding, urging them forward by appealing to their patriotism could win an economic war in the same way as a military one, he believed. The Conservative government studied the idea on a number of occasions; they came to the conclusion that the Trade Unions could never be persuaded that it was not a hoax to exploit their members by holding down wages with no reciprocal guarantee of increased purchasing power.

'You fear the trade unions,' Ferguson wrote to R. A. Butler when he was Chancellor of the Exchequer. 'Let me assure you that such a fear is quite unjustified. I have been closely in touch with the leaders of the trades unions and hundreds of work-people for the past two years and I have not found one person who is not in full agreement with the idea of raising the standard of living by bringing down prices. They now realise that the standard of living will not be raised by increasing wages.'

Butler was peeved. 'You make bold to state that I "fear the trade unions". That is not so. What I have done is to try to gauge their reaction to the proposal you advocate. In my judgement, the trade unions would perceive, and would resent, the fact that such a campaign by the government offered no certainty of increases in real purchasing power and some risk of the reverse both at home and in export markets.'

Thus Ferguson did not make much progress in having his scheme adopted, but this did not prevent him from continuing to badger successive Conservative ministers and prime ministers. And he also sent his lengthy letters to any and every statesman or other person in positions of influence whom he considered had the qualities of leadership necessary to put his scheme to work. Robert Menzies, Dean Acheson, Truman, Paul Hoffman, Eisenhower, Diefenbaker, and many others received his epistles. When Russia declared a succession of reductions in retail food prices in the early fifties, he was more than ever convinced that Communism was craftily manœuvring its way towards an economic victory that would be far more devastating than an armed war. So certain was he of the rightness of his views that he even offered a prize of £10,000 to the company or individual making the greatest contribution to price reducing.

To trained economists, Ferguson's proposals were almost embarrassingly naïve and it made them uncomfortable to hear them propounded as a great revelation of truth. On the other hand, in recent years, every government of economically powerful countries has taken a stand against inflation. The true menace of rising incomes and prices has been increasingly recognised. Only a very few years ago, whenever inflation threatened, stock markets boomed as investors bought common shares on the assumption that their prices would rise at least in step with other prices and that the purchasing power of their capital would therefore be preserved. Today, however, talk of inflation usually spreads despondency in the stock markets, for investors have come to expect governments to crack down with anti-inflationary measures of such severity that companies find their expansion slowed and their profits under pressure. There has been a marked change in attitudes towards inflation and it is treated with less tolerance than in the past. In his way, therefore, Ferguson was again ahead of his time. If his ideas could be sneered at as impracticable and oversimplified, they also contained a firm basis of truth.

Even if Churchill and other politicians bombarded by Ferguson were baffled by how to implement a Price Reducing Scheme, and even if they were irritated by Ferguson's inability to lay down well defined tactics, they still recognised this element of truth in the proposals and were impressed by the unassailable sincerity of the man making them. It is not surprising that this, coupled with his contribution to Britain's economy, caused Churchill to propose that Ferguson should be offered a knighthood. In November 1953 Ferguson received a formal letter from 10 Downing Street asking whether 'this mark of Her Majesty's favour would be agreeable' to him.

Ferguson's reply is interesting and what follows is the main body of his letter. 'I am greatly touched by the Prime Minister's thought. I most deeply appreciate his kindness. I wish I were more worthy of such a recommendation.

'I feel I should give you the reasons why I should not accept this honour.

'For great and brilliant statesmen like Sir Winston, who serve their country in the highest sense, no honour could be too great. Also for great soldiers, such as Lord Alexander, the Honours List is a splendid thing.

'I have the feeling, however, that the Honours List for industrialists may not be a good thing for the country. Unfortunately, many people in the industrial world do things, which they ought not to do, for financial gain. My experience is that the same kind of people will do even worse things when seeking to be included in an Honours List. It is substantially for this latter reason that I believe industrialists should not be recommended for Honours.

'Industrialists have the opportunity to become famous, if that is what they seek. They also have the opportunity to amass wealth and all that wealth can bring them. That should be sufficient for them!

'I fear I have seen, so often, the harmful effects of an Honours List for industrialists, that I believe I have come to the right decision when I ask that the Prime Minister would not submit my name to the Queen in this connection.'

Throughout his life, Ferguson only ever accepted two important honours: an honorary Doctorate of Science from Trinity College, Dublin, and later, in 1948, a similar degree from Queen's University, Belfast. He was particularly glad to receive the latter, for it came during the lawsuit against the Ford Motor Company when his ability and knowledge were under attack by the Ford lawyers. A stranger honour, and one that it was even stranger he should have found pleasing, came his way in October 1951 when the Abbot of a Trappist monastery in Belgium blessed Ferguson and his tractors. The Abbey of St Remy at Rochefort had just purchased a second Ferguson tractor and a rally was held to which many Fergusons owned by farmers in the region came to receive a blessing from the Abbot.

'At a time when factories all over the world are working day and night to produce war material and destruction machinery, it will be a real joy to me to call the celestial blessing upon these instruments of peace and work.

'May God, Who is not only the Lord of Armies but above all the Lord of Peace, bless these tractors which are a homage of the human genius to the superior Genius of the Creator.

'May He bless the inventor of the Ferguson System who spends the best of his time and his forces to create these mechanical marvels for the progress and well being of humanity.'

Harry Ferguson may have had an implacable hostility to the Roman Catholic Church, but nevertheless this blessing touched and pleased him.

27 *Merger*

Those who would cast doubt upon Ferguson's sincerity will find themselves dumbfounded by the fact that he applied the Price Reducing Scheme to his own business activities. He had never attempted to make much profit on the tractor, but he did have a reasonable margin on the implement range. With the passage of time, however, and as his convictions over the evils of inflation became deeper rooted, he increasingly attempted to reduce the price of his products, or at least hold them steady, by absorbing his suppliers' price increases. There were many periods, particularly in the late 1940s and early fifties, when the demand for Ferguson equipment was outrunning the supply. And there was never any difficulty in those years—except when the Argentinian government cancelled a large order because of foreign exchange difficulties—in selling all that could be produced. In these circumstances, a conventional pricing policy would have allowed the Ferguson companies to build up large cash reserves to protect them against difficult periods. Instead of this, Ferguson insisted on keeping prices down to the point where returns were needlessly small. His ambition was to reduce the price of the Coventry-built tractor from the figure of just under £200, at which it originally sold, to £100; and whenever he was ultimately forced to raise the price because production costs were increasing—to such an extent that he would have been forced out of business had he not done so—he bemoaned the fact at length.

Then in 1952, his Detroit-based company ran into difficulties. There was a general slackening in demand for farm machinery in North America, and all manufacturers found the conditions troublesome. Harry Ferguson Incorporated was worse placed than some of the others; its assembly plant was a relatively high-cost operation, and it was certainly not geared for a cut-throat price war in the marketplace with giants such as Ford. The result was that the profits made by Harry Ferguson Incorporated in the first years after building the assembly plant were eroded until the company was in danger of running a deficit. The vulnerability of the company made Ferguson fear that Fords might drop the price of their tractor so low—subsidising it from their other interests—that he would be forced out of business in North America.

There was yet another matter of great concern to him. By now he was approaching his seventieth year. His daughter Betty had recently married and he had no son in whose hands he could leave his industrial empire. Incidentally, it seems doubtful whether, even if he had had a son, he would have been able to assume such a responsibility. Although an affectionate father to Betty, he was also demanding and possessive. Ultimately she found the atmosphere so stifling that she decided to escape to Paris for a year. Had she not been endowed with some of her father's fighting spirit she might never have succeeded in freeing herself, for there was a lengthy struggle involved in doing so. Had Ferguson had a son, and had he been a fighter too, he might have left home or have been so repressed that his personality would have been permanently damaged. Such repression would not have been caused through deliberate unkindness, merely through the demands and intolerance of a father unaware of what he was doing to his son's psyche.

By the early 1950s Ferguson must have been disillusioned by the lack of progress made in promoting the Ferguson Plan. His hope had been that the post-war years would bring a universal attempt to mechanise agriculture. The unfeasibility of such mechanisation in developing countries has already been discussed in connection with the Bethesda demonstrations, but it should be reiterated that Ferguson always saw his equipment as completely utilitarian. It had a great task to perform in the world, and therefore it must be, and remain, as functional, and as cheap as possible. It was as near perfect as it needed to be, and therefore struggled against incorporating any changes on the grounds that they would increase the price, and make the provision of spare parts more complicated. He was also fearful of any changes that might make the equipment more demanding in its operation. It had to be easy to use if it was to give of its best the world over.

But, during the post-war years, the market for farm machinery did not open up in the developing countries. By far the greatest part of the machinery produced was sold in the regions of the world where agriculture was advanced and where industrialisation had taken place. The result was that, by the 1950s, the various tractor and implement manufacturers were left battling against each other in North America and Europe. In order to win a place in these sophisticated markets, it became necessary to make so-called improvements to tractors at regular intervals. Rather as in the automobile industry, the product had to be tampered with, the styling changed, and new features incorporated to improve its sales appeal. Inevitably costs rose and the machinery started to become so complex and sophisticated that only a well educated farmer would be able to make optimum use of it. This trend that began in the fifties is still evident today; in a sense it is a tragedy that more simple and inexpensive farm machinery is not available for the farmers the world over who are not able to write off the cost of their gleaming, beautifully-styled marvels of engineering against their tax bill. Ferguson was bitterly disappointed that his concept of functional and universal farm machinery, at low cost, had been submerged along the path of what was termed progress.

All the aforementioned factors—plus some new interests in the automobile field—contributed to make Ferguson believe that it would be expedient for him to hive off at least part of his farm machinery business. Most expedient would have been to dispose of Harry Ferguson Incorporated, the Detroit company whose losses were a drag on the much more successful Coventry company, Harry Ferguson Ltd. And this was the first tactic that Ferguson tried in the latter half of 1952.

Among the large farm machinery companies of the world, there was one which was likelier than the others to be interested in a deal with Ferguson. This was the old and respected Canadian firm of Massey-Harris. The company had grown out of the businesses begun by Daniel Massey and Alanson Harris, in 1847 and 1857 respectively. The two men, who had started out in a very small way, prospered greatly and even developed major export markets. Massey began by exhibiting his equipment at international fairs and exhibitions, starting as early as 1867 in Paris. This, and exhibitions at Antwerp and London, laid the groundwork of an overseas sales network, for importers from many countries made contact with the company at exhibitions and later established distributorships. By the 1880s both companies were selling their products the world over—except in the United States which raised a tariff wall—and had branch offices in a number of countries. There were able people in the Massey and Harris companies and it is not surprising therefore that they decided to merge rather than continue competing against each other.

Daniel Massey had based most of his expansion on buying rights to manufacture good products developed elsewhere, and the newly formed Massey-Harris Company did much the same until shortly before World War II when it developed the first good self-propelled combine harvester. This reinforced the company's position as the leading manufacturer of harvesting machinery. It was in this sphere that it excelled, and even though it did produce a complete range of farm equipment, by 1952 its tractors and tillage implements were outmoded and a weak link in its product line. Ferguson, on the other hand, was marketing the most advanced tractors and tillage equipment in the world, and hence there was an affinity between the companies.

The president of Massey-Harris was James S. Duncan, who had joined the company at the age of seventeen and worked with it in France, Canada, Germany and Argentina. Son of the original importer of Massey implements into France, an able linguist and a cultivated man, he was already General Manager of Massey-Harris by 1935, and President by 1944. He had done wonders in lifting Massey-Harris into its position of prominence following a difficult period during the Depression. His first contact with Harry Ferguson had been at Detroit at the time when Ferguson had been trying to find another manufacturer after the 1947 break with Ford. Duncan attended one of the promotional meetings that Ferguson held, but, like the rest of the potential manufacturers, he considered that Ferguson's future was under too black a cloud to warrant capital investment at that time.

Subsequently there were sporadic contacts in Britain between the two companies when Guy Bevan, the Vice-President of Engineering of Massey-Harris Ltd. met with Alan Bottwood, Ferguson's managing director, to discuss the possibility of engine supply and some other matters. By 1952 these contacts had taken on substance, for the Ferguson Company engineers were developing a combine harvester that could be mounted easily on the tractor and taken off again at the end of the harvest, thus obviating the need to have a power plant idle for perhaps fifty weeks in the year, as is the case with the conventional self-propelled combine. Massey-Harris had spare production capacity at its Kilmarnock plant in Scotland and was interested in manufacturing the mounted combine for Ferguson. The negotiations went ahead on the Bevan/Bottwood level until, by June 1953, agreement in outline had been reached on the manufacture of the combine and on the supply to Massey-Harris of the engine used in Ferguson tractors to power some of their harvesting machinery.

At that point Bevan asked Duncan to come to Britain to discuss the final details with Bottwood. The talks went smoothly and it was agreed that they would all meet again about a week later, on June 30th, to sign the necessary papers. But when Duncan and some of his staff arrived at the Coventry offices at the appointed time they found a very agitated and embarrassed Bottwood waiting for them.

'Mr Duncan,' he said, 'I feel terribly ashamed but we are not in a position to sign the contract. As you know, this company is owned, operated and dictated to by Mr Ferguson and when I called him in Ireland where he is convalescing from an operation he told me that he didn't agree with this thing or wish to continue with it. However, he does wish to see you before you leave England. He is coming back from Ireland immediately and he would be pleased if you would have lunch with him at Abbotswood. He has sent his car for you.'

In front of the office block stood Ferguson's Rolls-Royce with chauffeur in attendance. Duncan told Bottwood not to be upset by this sudden reversal and set out for Abbottswood, where he must have arrived with perfect punctuality, for he found both Ferguson and his wife awaiting him on the doorstep. The lunch was excellent, the host at his most charming, and after the meal in the beautiful dining room, Ferguson said: 'Well, Mr Duncan, I didn't ask you to come here just for lunch. If you'd care to come out to the summer house in the garden to take coffee I would like to explain why I cancelled the combine proposal.'

Once esconced in the summer house in the corner of that colourful and tranquil garden, he went on: 'This may surprise you, but the proposition you were making over the combine was so much smaller than what I have in mind that we should just forget it. And I'll tell you what I do have in mind.'

He continued by complimenting Duncan on his achievements in building up Massey-Harris, particularly in the export field, and then said: 'I have come to the conclusion that you could do more for our exports than we have been able to do and therefore I'm going to offer you a partnership in my company. You will need to put up no money but you will have to resign from your position with Massey-Harris. You'll get 50 per cent of my business.'

Duncan, very surprised, courteously refused to leave the company into which he 'had been born'; at this Ferguson became huffy.

'You'll regret it all your life. You'd make more money with me in two or three years than you probably will in the rest of your life. Still, if that is your view I have nothing more to say. Let's look around the garden, shall we?'

After a time spent strolling between the orderly shrubs and flower beds and talking about the difficulties facing the farm machinery industry in general, Ferguson said in his brisk, impulsive way: 'Let's go back to the summer house, Mister Duncan, and talk some more.' Once there, Ferguson pressed him as to whether the offer was absolutely unacceptable, and was told it was.

'Well, in that case, since you're not interested in a partnership how would it be if our companies got together? I might be willing to sell some of my interests to Massey-Harris.'

This was of immediate and intense interest to Duncan, aware as he was of the outstanding success of the Ferguson System, the excellence of its products, and the immense goodwill the company had created. He was also well aware, however, that even if Ferguson was not a businessman in the accepted sense of the word, he was not naïve enough to make a proposition if there was likely to be a loss of money to himself involved. So Duncan proceeded cautiously.

'That's a different matter,' he said. 'On what basis would you consider doing this?'

'Well, what I would like to do is sell your company my American business. Then perhaps little by little you could buy an increasing share in the British and world business, a certain percentage per year which we could agree on.'

'That's worth considering,' said Duncan, 'but, as far as I'm concerned, not very favourably. I know that your American business is in rather a dilapidated condition and I couldn't recommend to my directors to buy a company which is not prosperous and which shows no sign of becoming so. Nor would I recommend buying a proportion of your British company without having some say in its management.'

'Well, I think you're making a big mistake,' Ferguson replied, 'but I've done my best so there it is.' Before he left, however, Duncan was persuaded to go to Harry Ferguson Incorporated in Detroit to look the company over and discuss with D'Angelo a possible merging of interests.

Even if that first intimate meeting was inconclusive it established one important thing very firmly: Ferguson and Duncan liked each other, and it was particularly important that the former should like the latter if there was to be any future intercourse between them. About two weeks after his return to Canada, Duncan and two of his top management journeyed to Detroit to examine the operations of Harry Ferguson Incorporated. They were given access to everything they wanted to see and were treated with the greatest courtesy by D'Angelo and Hermann Klemm, the chief engineer for Harry Ferguson Incorporated. It is interesting that Ferguson had not at this stage told his Detroit team the full implication of his talk with Duncan. They knew that Ferguson was beginning to feel the burden of his responsibilities and did not consider he had the time or energy to help his management with commercial problems. In particular, they believed that the main objective of the contacts with Massey-Harris was to attempt to share distribution outlets, for after the break with Ford, the Ferguson distribution network had never risen to its previous strength.

Duncan and his men were greatly interested by some new implements that the Ferguson company had developed. A few days later Duncan wrote in a memorandum: 'I place great stress on the engineering quality of the Ferguson-designed implements.

This all stems from Harry Ferguson's genius and may never be duplicated by Ford, just as after eighteen years it has not been successfully duplicated by anyone in the orthodox implement industry.

'Basically we must ask ourselves if the trend is towards the Ferguson concept or the old line implement company concept of engineering. If it is the former—and I believe it is—then the value of an association with its originator might be very great and might place us in a position of leadership in the industry and should, within the next few years, add at least another 100 million dollars to our turnover.'

But the Massey-Harris directors were only interested in an 'association' if it were to be world-wide. Their company stood to gain a great deal on such a basis, whereas a link-up in North America only would have been a shakier proposition.

Duncan wrote to Ferguson on July 16th, 1953, saying that he was convinced after three days at Detroit that the two companies should 'co-ordinate their efforts in North America' provided that this was viewed as a 'first step towards a still closer union' of their two organisations as a whole. He suggested coming to England again on August 4th to discuss the matter with Ferguson. Ferguson wrote back accepting the suggested date and inviting Duncan and his wife to stay at Abbotswood.

Crucial in Ferguson's mind to the whole outcome of the negotiations with Massey-Harris was a new tractor—hitherto unmentioned in this book—that his engineers had developed. The idea of producing a larger version of the Ferguson System tractor went back to the war years with Ford when a good deal of experimental work was done without ever reaching the stage of a production prototype. When the Coventry-based company came into being after the war the 'big Fergie' was a top priority for the Engineering Department. By the time of the negotiations with Massey-Harris, some so-called LTX prototypes had been built and tested. The idea was to market it as the TE 60 and it was in fact basically an enlarged version of the TE 20.

Those who had the privilege to work with the LTX and carry out its rigorous field testing programme still speak with awe of it. It was by all accounts a phenomenal tractor and although today enormous wheeled Ferguson System tractors pull five and more furrow ploughs with ease, the LTX pulling such a plough through the stiffest clay the company could find—near Southam, south of Coventry—was vastly exciting to the enthusiastic and almost fanatical Ferguson staff. The still-live sense of marvel among many of those who witnessed the LTX tramping remorselessly forward in the toughest of conditions probably stems from the fact that they had never before seen any wheeled tractor with such power and wheel grip. Today's large Ferguson System tractors are doubtless better, but nostalgia for the LTX and the impression made by that first 'big Fergie' linger on among the old Ferguson engineers.

A demonstration of the LTX was to be the linchpin of Harry Ferguson's attempt to snare, once and for all, the interest of Massey-Harris. The last part of the letter which he wrote to Duncan when inviting him to visit Abbotswood on August 4th, 1953, was a minor masterpiece in setting the scene for the talks and the demonstrations:

'If we have a satisfactory talk then we will proceed at once with your technical investigation and the necessary demonstrations. The only qualm is that we may not be able to get conditions sufficiently difficult to show the merits of our new big tractor, in particular because the harvest will not then be gathered and all the land is occupied. However, we will do our best.

'I note you have some commitments on the continent. I am sure, however, that you would not think of leaving any technical investigations to others without being present yourself. It is these technical investigations which will create that necessary confidence between us which could not be created in any other way.

'A company can be no greater than its products and I know that we have something of immense value to put before you for the whole future of your company.

'Thank you so much for your very nice letter of July 3rd. You are a nice gentleman and I do look forward to seeing you again.'

Prior to leaving for England, Duncan and his board of directors discussed every possible permutation for a deal with Ferguson in order to be ready for any eventuality that might arise. They did not know what Ferguson might be prepared to offer, but he had indicated in a letter that he was willing to take a step towards them. On the occasion of their previous meeting, Ferguson had told Duncan that he wanted to talk business only with him, and not with the management of Massey-Harris generally. Nevertheless, Duncan felt he had to take Harry Metcalfe, his comptroller, with him and on reaching London Airport he telephoned Abbotswood to ask whether Metcalfe could join them. Ferguson raised no objection, because D'Angelo was also to be present, but the luncheon that he gave at Abbotswood upon Duncan's arrival was for Duncan and his wife only: Metcalfe and D'Angelo were to eat elsewhere, for Ferguson had no inclination towards making undue fuss of the generals when the field marshals were present.

After lunch Ferguson suggested a chat in the summer house where, with a minimum of preamble, he said, 'Well, I'm prepared to sell you the whole company. What conditions would you have?'

This was exactly what Duncan and his management had been wanting but had hardly dared hoped for. Nevertheless he had prepared himself in case the offer was made.

'I can tell you vaguely,' Duncan said, 'but I think probably this should be done with a little more formality and with some of our people here.'

'No, no,' Ferguson said. 'A gentlemen's agreement is quite good enough for me. Sit down now and tell me your conditions.'

Duncan duly outlined their offer, which involved a swap of Massey-Harris shares; Ferguson noted the details, as Duncan recalls, on the back of an envelope, but it would be surprising if it were not in his ever-present notebook.

'That seems fair. I agree with it,' Ferguson said when Duncan had finished. 'I'll sell you my company on that basis.'

Duncan was astonished that agreement had been so quickly forthcoming. He would have been prepared to offer more if necessary, aware as he was that he was buying tremendous goodwill, very superior engineering knowledge, a company of the highest prestige, well known throughout the world, and a company which in many areas had complete control of the tractor business.

By this time D'Angelo, Metcalfe and Bevan had joined Ferguson and Duncan. D'Angelo sat in silence throughout. He was the first of many Ferguson employees to be sick at heart over the sell-out of the company. Working for it had been the most intense experience of their lives; they had helped build it into what it was and the loss of immediate control of the company—even if they had always been subject to the dictatorial powers of the 'Old Man' —was like the amputation of a limb.

Knowing, in fact, how much resistance and sadness there would probably be, Ferguson had not even taken Alan Bottwood, his managing director at Coventry, into the full confidence of his intentions. Bottwood would have fought strongly against any complete sell out, and since this was certainly not the moment for internecine squabbles, Ferguson went to the length of asking Duncan to say nothing to anyone; and he sent Bottwood, in ignorance, to Wales to see Sir John Black during the following week-end when a group of Massey-Harris directors arrived and the final merger details were settled. It was a cruel thing to do, and one which it is no pleasure to record here; it was an action unworthy of Harry Ferguson. Many people have commented that the sale of the company, in such circumstances, broke Bottwood's heart.

'That's all settled then,' said Ferguson after details of the sale had been gone over with Metcalfe, D'Angelo and Bevan in the summer house. 'Oh no, not quite settled,' he added, 'because there are just two points I would like to raise and which I hope you'll be able to accept. I'm very proud of what I have been able to do . . .'

'Quite rightly,' interjected Duncan, 'you've done a great thing.'

'. . . and I would like to have some position of honour in your company. My pride is involved in this, and the other thing I'd like to ask is that it be presented as an amalgamation of our companies rather than a sale. I'd feel better about it.'

'I couldn't agree with you more,' replied Duncan. 'I'm president and chairman of the company in Canada, and chairman is an honorary rather than an executive position in North America. It's where presidents go when they are kicked upstairs. Your function would be to come to our annual meeting and take the chair.'

'You mean you'd make me chairman of your company,' asked Ferguson, 'and present the sale as an amalgamation.'

'Yes, of course,' said Duncan.

Ferguson pulled a handkerchief from his pocket and wiped two tears from his cheeks. 'Mr Duncan,' he said, 'I knew from the time I met you that you were the man I wanted to be associated with.'

Somewhat embarrassed, Duncan dismissed the matter as inconsequential.

What with interruptions for tea and other matters, it was now approaching six-thirty in the evening.

'Now, Mr Duncan,' said Ferguson, 'dinner is at seven o'clock so you're in nice time to get dressed.'

'Well, look, I want to phone Canada to let the chairman of my executive committee know what I've done. I think it's only fair. And I'd like to have a bath because I travelled overnight on the plane, so would you mind making it, er, half-past seven for instance?'

Absolute consternation swept over Ferguson's face. 'You know I always dine at the same hour,' he said when he had overcome his shock. 'But I suppose this is something of a special occasion after all. I'll go and talk to Maureen about it.' And he trotted away up the garden.

Duncan lingered for a moment to talk with his comptroller and then made his way towards the house. He met Ferguson coming out. 'I've spoken to Maureen,' he said, '. . . but dinner is at seven o'clock.'

In a skelter, Duncan managed to make his telephone call to his delighted colleagues in Canada, scramble into his dinner jacket and rush downstairs with his wife. Ferguson in his customary velvet dinner jacket was waiting with Maureen to offer them sherry. At precisely seven p.m. the dignified butler appeared to announce that there had been a hitch in the kitchen and dinner would be delayed. Duncan had the greatest difficulty concealing his amusement; Ferguson made no attempt to hide his annoyance.

A couple of days later, when the rest of the Massey-Harris top management was on its way to Britain to see a demonstration of the Ferguson equipment and to go through the formalities of the merger (and when Bottwood had been despatched to Wales), Harry Ferguson suddenly said to Duncan, with whom he was now on the warmest of terms:

'You know, Jimmy, I've been thinking this over. I was awake last night. I didn't sleep very well. I've had a lot of trouble with my engineers always trying to change things on my tractor. I believe in keeping it as it is . . . perfect as it is. You're going to be up against the same trouble. You are not an engineer and you're not acquainted with things the way I am and you're going to have a lot of difficulty. Now I'm going to make the suggestion that it would be very wise if we included in the arrangement we draw up in writing that all engineering proposals come to me and I have control of the design of Ferguson equipment. You'll find this will give you great strength because when they come to you and want to change this and change that you'll submit it to me and I'll be responsible for deciding.'

This eventuality of design control had been foreseen by the Massey-Harris management in their planning prior to Duncan and Metcalfe's departure for Britain, and therefore Duncan agreed to Ferguson's suggestion. In doing so he made what he later considered to have been his second mistake—the first having been to give Ferguson the position of chairman.

When the group of Massey-Harris directors arrived in England, they were treated to a demonstration of Ferguson equipment, including the LTX tractor and its implements, at Barford, a village near Warwick. We shall examine their opinion of the LTX in a later chapter, for a more important point still at issue when the Massey-Harris directors arrived was that of a million dollars. The book value of the Ferguson companies had been estimated by Massey-Harris at $16 million net at the end of 1952 and (as Duncan remembers it) Ferguson had given his outline agreement to using book value when assessing the number of M-H shares he would receive; but it was suggested by one of his advisers that he should ask for more than the $16 million figure as this was not the true asset value of the companies. In any event everything had been agreed with the exception that Ferguson wanted M-H shares worth $17 million.

The negotiations were stuck at that point until Ferguson, Eric Phillips, 'Bud' McDougald and Duncan, all of the Massey-Harris board of directors, were driving to another demonstration and had stopped in Broadway, a lovely Cotswold village. A good deal of inconclusive bargaining had been going on in the car.

'Gentlemen,' Ferguson suddenly said. 'Let's stop haggling about it. In Ireland we have a way of settling such matters. I suggest we toss for it.'

There was a brief moment of consternation among the Massey-Harris team at the thought of settling the matter of a million dollars, that were not theirs in the first place, by the spin of a coin.

'Go on,' urged Ferguson. 'I always lose anyway.'

Phillips was the first to recover and agree. He was something of a gambler by nature and in any case the book value of the Ferguson companies was bound to be a nebulous figure: it is difficult to value goodwill and engineering expertise with precision. Thus, in the street outside the Lygon Arms at Broadway, Ferguson turned to his financial adviser, John Turner, and asked him for half a crown. The coin was spun and Ferguson called tails . . . he lost. This prompted him to say, 'I will toss you for the coin, and that is the sort of bet I win', whereupon it was once again tossed and Ferguson won. The coin was about to be returned to Turner when McDougald intervened. Subsequently he had the coin mounted on a silver cigar box with the words 'The $1,000,000 Coin' engraved under it. Also inscribed on the box was the dedication 'To our friend and partner Harry Ferguson. A gallant sportsman.'

Thereafter the merger was merely a question of formalities, perhaps the most difficult of which was the drafting and frequent re-drafting of the press release that announced it and explained the future operation of the two companies under the name Massey-Harris-Ferguson Limited. The keynote of the release was one of 'pooling their organisations which are spread wide over the earth'. The release of course dwelt also on the Ferguson System that had 'blazed a new trail throughout the world' and on the fact that the Massey-Harris Company had 'pioneered the self-propelled combine in every country where wheat is grown'. Thus the amalgamation of the two companies was 'probably the most important news in the farm equipment industry in the present century'. We can guess who was responsible for making sure that that last phrase was included.

Harry Ferguson received 1,805,055 Massey-Harris shares worth about $16 million (£5·7 million) at the current price of just under $9 (£3.25) each. The new farm machinery company of

Massey-Harris-Ferguson was among the world's largest in terms of sales volume, and Harry Ferguson was the largest single shareholder. As we shall see, the new company ran into difficult times and Ferguson himself was to leave his position of Chairman after a year wrought with tensions and discords.

The news of the merger came as a stunning shock to all Ferguson staff, not just to Alan Bottwood. Two Ferguson directors were appointed to the board of the new company, but this did little to mollify the feelings of the ex-Ferguson employees. During the years following the merger and while the new company was in the throes of painful reorganisation, there was some acid rivalry between the old Massey-Harris 'reds' and the Ferguson 'greys'. Comments such as 'The Old Man has sold us down the river,' were commonplace among ex-Ferguson staff until the company began to assume its new identity and purpose, as well as its unified product line and its name of Massey-Ferguson. In the difficult period Bottwood, D'Angelo and many others left.

Regarding the name of the company after the merger, Harry Ferguson once joked that the announcement of the amalgamation should have been made in the deep south of the United States. When asked why, he replied, 'Because there the company would have been called "Massa'-Harri'-Ferguson".'

28 *Of Cars and other Matters*

To have left such a large mark on the world of farm machinery would have satisfied most men. However, Ferguson's involvement with tractors and ploughs was one that came about almost accidentally during World War I, as we have seen. His interest prior to that, and his great love, was the automobile to which he had already made a contribution through his expertise with carburettors. And, as a much younger man, in the days when he had been involved in the Ulster T.T., he had often propounded ideas on improving automotive design. He was so earnest and insistent in his buttonholing of members of racing teams and in offering them advice, or in criticising their existing designs, that it often took all their tolerance and humour not to ask him to go and talk to somebody else.

It is quite certain that it would have pleased Ferguson even more than did his farm machinery success had he been able to make a really important contribution to automobile design. (Henry Ford I was just the opposite: despite his enormous success with cars, his heart was largely in the farm tractor.) Therefore, through all the years of work and achievement with the Ferguson System, its inventor still carried a passionate interest in cars.

One of the great figures of pre-war racing with whom Ferguson had become acquainted was Freddy Dixon. Dixon was a wizard at tuning Riley cars and then at driving them with a flair that allowed him to keep up with, and often beat, cars with much bigger engines. One of the great sights of pre-war motor racing was Dixon in his Riley cheekily subjugating the more expensive, works-backed cars. His successes were of irresistible appeal to many who appreciated the comic and the David-and-Goliath element. Ferguson was among these admirers and was delighted when Dixon repeatedly used his garage in Belfast as headquarters for his participation in the Ulster T.T. During these occasions the two men often discussed the possibility of jointly developing a better automobile, but Ferguson was too involved with his tractor interests to follow it up.

Also an admirer of Dixon's—indeed as a boy he almost hero-worshipped him—was Tony Rolt, a young racing driver who took up the sport seriously when he bought the famous E.R.A. Remus from Prince Birabongse in the late thirties. Dixon had by then retired from active racing and Rolt went to ask him for his help in tuning and maintaining Remus. The two men became friends and Dixon unveiled his plans to build a car for an attempt on the world land-speed record. He envisaged a number of advanced features such as four-wheel drive, all independent suspension and inboard brakes. Rolt suggested that it would be more useful to try to apply such features to a vehicle for road use, and offered to help do so. Dixon agreed and a company called Dixon-Rolt Developments was formed. By the time World War II broke out, they had designed and started to build a strange vehicle named the Crab; they used the Riley engine out of Remus mounted on a tubular chasis with four-wheel drive and four-wheel steering that was activated by swinging the whole of the axle shafts rather than by pivoting the wheels alone. In 1939, even though it was not finished, Rolt tried to interest the Army in the Crab for cross-country purposes; he was a Regular Army officer, so he had the right contacts, but it was not an auspicious moment for introducing new ideas.

During the rout of the British forces in France in 1940, Rolt was taken prisoner at Calais. He was obviously a great nuisance to his captors, for he was finally put into Colditz, the castle where the most persistent would-be escapers were kept under maximum security. When liberated, Rolt and a small group of other prisoners had just completed building, out of all manner of odds and ends including sawn up floorboards, a very professional looking glider which they planned to launch from the Colditz ramparts.

Dixon-Rolt Developments and the Crab had languished during the war years, but Rolt got back to work on it at once, often helped by fellow ex-prisoners to whom, during those long years spent in captivity, he had enthused about the project.

At this point, Harry Ferguson returned to Britain from Detroit, and high on the list of people to look up was Freddy Dixon. For

Ferguson still held an idea of attempting to improve motor car design. Dixon and Rolt went to see him at Broadway, where he was staying prior to buying and moving into Abbotswood. Ferguson and Dixon had been in spasmodic contact during the war, but Ferguson was not acquainted with the Crab project. When he heard about it he expressed much interest. 'Keep going, boys,' he said. 'Some time we'll have a talk about it.'

Over the following two years Ferguson gave some funds to Dixon and Rolt. The amounts were not large but they enabled wage and material bills to be met and thus ensured continuity of work on the Crab. Then, in 1949, Ferguson paid a visit to the Dixon-Rolt headquarters in Redhill to see a demonstration of the Crab. The powerful, lightweight four-wheel-drive chassis was amazing to watch in action: it scuttled up and down very steep slopes with astonishing agility. Ferguson was much intrigued, for it seemed to him that here lay an opportunity which could be developed into a novel concept of automotive engineering. By coincidence, at that precise time Dixon and Rolt were seeking more finance and they valued the possibility of obtaining it from someone as enthusiastic and knowledgeable as Ferguson. However, an arrangement with him was not to be as easy as they had hoped. He wanted complete control or nothing, but Dixon, as a rugged individualist, did not want to be subservient to anyone; he hoped Ferguson would merely provide the means for the development work.

However, the Ferguson lawyers drove a hard bargain in their offer for the patents and assets held by Dixon-Rolt Developments. Ferguson's attitude was that by taking over the financing of the project he would be allowing Dixon and Rolt to pursue their work unhampered by worries about money, and if the project was successful they would all become millionaires. 'A typical Ferguson approach' in Rolt's words. 'Nothing today, but jam tomorrow!' (Rolt never compared notes with Sands but it is remarkable how similar is his comment to many made by Sands.) Rolt appreciated, however, that the backing of someone like Ferguson was essential and he submitted with good grace to the conditions imposed. Dixon submitted too, but with less grace, and a new company called Harry Ferguson Research was launched in April 1950, with considerable publicity. The press was agog over the announcement of the founding of Harry Ferguson Research. For by then his genius was recognised by many and his lawsuit against Ford had made him well known. There was much speculation as to what might be expected from his active interest in automotive research.

Ferguson did not only want complete control: he wanted to take an active interest in the development programme, and he began to lay down how things were to be done. Dixon found this irksome and the first row was not long in coming. One source of contention was Dixon's rumbustious and hearty approach to alcohol. Ferguson deplored his drinking habits, which had caused brushes with the law on a few occasions, and he disapproved of the occasional sprees he indulged in. Neither did he hesitate to tell Dixon so, but Dixon did not even consider that he drank to excess and understandably wanted to know what business it was of Ferguson's, even if he did. Gradually Dixon began to take less and less part in the development work. He was perhaps piqued also by the activity and competence of an engineer called Claude Hill who had joined the team shortly before the founding of Harry Ferguson Research. Hill had been with Aston Martin as their Technical Director and he had designed the prototype car which was marketed as the DB2, after David Brown bought the company. The DB2 was the forerunner of the series of prestigious cars that have since borne the name Aston Martin. Whereas Dixon was a self-made engineer who worked from empirical knowledge and hunches—and perpetrated some disastrous errors as well as achieving great things—Hill brought to bear a wealth of skill and fundamental knowledge that only sound training could impart. Several of the Crab features were scrapped as impractical, notably the four-wheel steering by swinging axle shafts. It proved impossible to steer the vehicle under braking, and tyre scrub was excessive.

The plan was still to produce a revolutionary family car of unprecedented safety and roadworthiness. It was to have 100 per cent wheel grip on all surfaces, and stability, cornering and braking, superior to the best modern sports car. Among its novel features were to be: a single disc brake acting on the transmission to slow all four wheels evenly and minimise the risk of skidding; all independent suspension; and the driver centrally positioned behind a wrap-round windscreen with passengers to each side and slightly behind him. But of prime importance was to be its four-wheel drive, and here the system on the Crab proved inadequate; for it had a free wheel arrangement to allow the front wheels to overrun the rear, as necessary, and when they did so the front wheels were not being driven. Ferguson wanted four-wheel drive at all times.

In addition to designing a new gear box and engine for the Crab, Hill wrestled continuously with the problem of constant four-wheel drive through a controlled differential. This was the toughest nut to crack, and upon it depended many other elements. What was necessary, as Ferguson once put it, was 'a diff that diffed when it was supposed to and didn't when it wasn't'. Put in less cryptic terms, a differential was needed that would fulfil its classic task of allowing each of the wheels to rotate at a different speed from the others when necessary (to cater for different effective rolling radii, and for cornering) yet at the same time give positive drive to all four wheels and not permit one or two wheels to spin independently of the others. The problem was one that many engineers had tried to solve before, without success. Hill worried at it from all sides and lost a good deal of sleep in the process. Finally one night, after his wife had gone to bed and he was sitting downstairs at 2 a.m. pondering, the solution came to him. Clearly and without excitement, the basis for a controlled differential unfolded in his mind. The only paper he had available was the blank 'stop press' space of the *Evening Standard.* He sketched out the rudiments of his idea and calmly went to bed. The next morning, after sleeping soundly, he made a proper drawing of his idea and showed it to Rolt. Rolt happily agreed that the solution had indeed been found.

There was to be a meeting with Ferguson a few days later so they decided to wait until then before telling him of the invention. When they arrived they found that Hermann Klemm, the chief tractor engineer from Detroit, was already in the summer house with Ferguson. They walked across the lawn to join them. Hill was very conscious of the drawing in his pocket, but being a quiet and unassuming man he was not going to show his hand until asked. Klemm, a clever engineer but stolidly Teutonic in his attitudes, was less reluctant to speak up. Ferguson asked him for his views on four-wheel drive possibilities and Klemm began to pontificate. One needed to understand the problem fully first, he said, and the best means of obtaining fundamental knowledge would be by fitting strain gauges on axle shafts and carrying out a thorough series of tests. He continued at great length to explain his point of view; the drawing in Hill's pocket was shouting for attention. Finally Ferguson tired of the pedantic discourse.

'Yes, yes, Hermann,' he snapped, 'we know we have a problem and we know what it is. We want to know how to overcome it.'

'There's something we'd like to show you, Mr Ferguson,' interjected Rolt. Hill produced the drawing and Ferguson studied the layout of the controlled differential with its complicated arrangement of clutches, freewheels and planetary gears for a few minutes. 'This is it!' he said at last, looking up with a smile.

Not long after the excitement of the breakthrough on the controlled differential, Ferguson told his automobile research team that he wanted to move them to the headquarters of the tractor business in Coventry. Dixon and Rolt had come to Abbotswood to see Ferguson on a routine visit when he informed them. Dixon, who had indulged himself the previous evening and was unusually morose, was furious. A blistering row broke out during which Ferguson enumerated Dixon's transgressions like a court indictment. Dixon bluntly expressed his total disenchantment with the whole arrangement and said he wanted to leave. Ferguson replied that he thought it would be better if he did, and thus the association between them terminated. Rolt, Hill and the rest of the research team moved to Coventry and carried on their work in part of the tractor engineering premises.

Rolt was happy to be in Coventry because at the time he was driving for the Jaguar works team and the Ferguson and Jaguar organisations were only a few miles apart. Ferguson viewed Rolt's participation in motor sport with mixed feelings: he was passionately interested in it and delighted that Rolt was winning races, including important ones such as the Le Mans 24-hour event; but at the same time he was worried that the managing director of his research operation would kill himself. He fretted and wrote contradictory and voluminous letters to Rolt, but in the final showdowns his love of racing won and he never prevented Rolt from participating.

Coinciding with the move to Coventry and Dixon's withdrawal, Rolt and Hill decided that some of the ideas that had been in the development programme would have to be shelved. They wrote a memorandum to Ferguson explaining the need to rethink many aspects of the design. The central steering position was to be dropped because of likely sales resistance; the mounting of the engine and gear box on the single tube chassis was difficult if running silence and smoothness were to be obtained, and therefore the chassis should be redesigned; the engine should be in the front rather than the back. The list of fundamental changes was a long one and it brought the whole programme close to collapse. Ferguson was downcast by the number of changes that needed to be made. The team had spent over eight years on the Crab and to discard so much of the original concept, in addition to the earlier scrapping of swing axle steering and the first four-wheel drive system, was a severe setback. Ferguson's discouragement was shortlived, however, for when had he ever refused a challenge? The research went on, even if Ferguson irritated the team by frequently bringing other engineers to look at their work and offer advice. Perhaps he had temporarily lost some confidence in his staff, but whatever the reason it was a tactless thing to do.

29 *Sell-out*

'You'll quarrel with him inside three months,' prophesied Hennessey when Duncan, an old friend, was telling him about the merger of the Massey-Harris and Ferguson companies.

'Oh no, I get along just fine with Harry,' protested Duncan.

However, Hennessey's forecast came true—in half the time—for Duncan had underestimated the implications when he agreed to Ferguson being chairman of the new company and having complete authority on design matters connected with Ferguson equipment. He was under the impression that Ferguson really did want to give all his attention to automotive research and merely preside at the annual general meeting of Massey-Harris-Ferguson. The awakening was not long in coming: on September 23rd, about a month after the merger, Duncan went to Abbotswood again, and on entering found Ferguson standing tautly in the magnificent hall.

'What is this I hear, Jimmy?' he asked imperiously without wasting time over a greeting. 'I believe you have laid on a distributors' meeting here in Britain. Why haven't I been invited to attend?'

'Of course you haven't, Harry. We have distributor meetings all the time and there is no point in your coming. You are only going to preside at the A.G.M.'

'Oh no! Oh no! Not at all. I want to come to all your meetings. I'm Chairman of the company, don't forget.'

'You'll be doing an awful lot of travelling then, Harry,' Duncan rejoined with a laugh, and managed to change the subject.

In any case, Ferguson had something much more important to discuss: future pricing policies, on which he had prepared three pages of single-spaced typed notes for his talk with Duncan. His concern over prices on this occasion was not that of his Price Reducing Scheme; he was simply worried, and justifiably, by the high cost of the tractor as supplied by the Standard Motor Company. He believed that as long as prices were kept high they constituted an open invitation for other manufacturers to step in and undercut them. In particular, Ford could easily do so, he argued, perhaps even selling at breakeven prices for a couple of years and stealing the lion's share of the market. (It must not be forgotten that some important Ferguson patents were expiring, or shortly to expire, around this period.) This question of prices was one that had bothered him even before the merger with Massey-Harris and he tried to obtain a price cut from Sir John Black, who had tacitly admitted to others that his profit on the tractor was almost embarrassing. In fact the Standard Motor Company's post-war success was owed largely to the tractor production. It is significant that, with only a minute profit being taken by Ferguson, his tractor was being sold at 1s. 3d. (17½ U.S. cents) more per pound weight than the Fordson, whereas Standard Motor Company cars were selling at only 1½d. (1¾ U.S. cents) more per pound than their Ford equivalent. And this was so when the Standard Motor Company tractor plant was probably the most modern in the world and when its output was from two to three times that of the Ford tractor plant at Dagenham.

Ferguson, however, seemed unable to impress his concern on Duncan. Indeed, he later wrote, 'I have done my best with Mr Duncan but his response has been that of a man who is so satisfied with what is happening today he doesn't need to bother about the future.' (This was an astute and apt assessment, for it was an inability to keep pace with events, a lack of aggressive forward planning, that tumbled Duncan from his position of power in the company only a few years later.) So strongly and sincerely did Ferguson feel over the question of tractor pricing that he told Duncan he was willing to forgo all dividends on his large holding of Massey-Harris-Ferguson shares for one, or even two, years in order to be able to reduce costs to a point where other manufacturers could not undercut.

The kernel of discord in the question of pricing was a £20 ($56) reduction granted by Sir John Black as a result of Ferguson's pre-merger representations followed later by similar pleas from the Canadian management. It had been Ferguson's expressed intention to pass this price cut straight to the farmer, but Duncan decided that it would be more expedient to retain the £20 in the Company. Black was displeased and Ferguson was furious; he saw it as a short-sighted policy and continued to urge Duncan to reduce the tractor retail price.

The engineering and design of Ferguson equipment caused even greater difficulties between Duncan and Ferguson; the first and most serious trouble arose over the LTX or TE60 tractor. The outstanding performance of this tractor has already been described, but it did have a design shortcoming in that no tricycle version had been definitely planned for the American corn belt. Immediately after the merger, the Canadian management was in a quandary over whether or not to produce the TE60. Duncan requested a number of design changes, including the installation of a range of stock engines made by Perkins and Continental instead of the superior power plant designed for it specially by Ferguson engineers. He was seeking ways of reducing the capital outlay for manufacture, if they decided to go ahead. The installation of the Perkins diesel unit made the conversion to a three-wheel model even more difficult; and fundamentally, as has already been mentioned, Ferguson thoroughly disapproved of tricycle tractors anyway.

Ferguson considered the delay over putting the TE60 onto the market as procrastination and a sinful waste of a great opportunity. But if his feelings were running high, so were those of the engineers in Coventry and Detroit. The British team, led by John Chambers, had developed the TE60 and naturally wanted to see their brainchild in production. However, the Detroit team under Hermann Klemm were putting the finishing touches on a new and more powerful version of the small 'Fergie', the FE 35, for production in North America; and it seemed to the Coventry engineers that their Detroit colleagues were promoting the FE 35 to the detriment of the TE60. They deeply resented the precedence being given to Klemm's team, few of whom had the long experience of the Ferguson System that they had. It smacked of chauvinism to them—North American management favouring Detroit engineering. Their bitterness spread among non-engineering staff in Coventry too. The recently merged company was merged in name only; in spirit it was rift asunder. It had not merged physically either, for there were still separate dealer networks and product lines for Massey-Harris and Ferguson, the 'red' and the 'grey'.

Finally, in March 1954, the management decided on a compromise course of action for the TE60: they would manufacture a tractor similar in basic design to the TE60 but it would be marketed by the Massey-Harris dealer network. When this news reached Ferguson he did not seem to absorb it fully, but a week later he had and he was deeply disturbed. That Ferguson engineers were to assist in the design of a Massey-Harris tractor was a particularly grievous point, for he considered, with justification, that the Massey-Harris tractors of that era were much inferior and should be dropped from the company's product lines in any case. On April 20th, 1954, he wrote Duncan a fifteen-page letter protesting at the way the company was being run.

'I very much regret to tell you that the way things are heading in Massey-Harris-Ferguson has become so serious that I feel I shall never recover unless I can get relief from my anxieties . . .' his letter began. He went on to cover the matter of pricing policies yet again, urging the passing on of the £20 price cut to the farmer and telling Duncan to insist on further cuts from the Standard Motor Company.

'I'm sorry, Jimmy, but I can only describe what you say as being untrue and that your words are wild and irresponsible,' he wrote in refuting Duncan's allegation that the TE60 had been built for the Eastern Hemisphere and was unsuitable for North America. He claimed that the TE60 had been conceived by his team when they were with Ford in Detroit and that in 1953 a tricycle version was to have been designed and built in Detroit, without any difficulties then being foreseen, using the Ferguson-designed engine.

'Your trouble, Jimmy, is that you are being pulled this way and that by criticism of our machine after it has been mutilated by suggestions for changes to its design. . . . In 1939 when the late Henry Ford and we launched the Ford/Ferguson I was possibly the most laughed at and criticised man in the world. All the things you have just written me were said about me and they bothered me not in the least because I believed I knew my job.

'I have had so much criticism and hard things said about me that I would like to comfort you by telling you that what *you* have said has not made me angry.

'At this point I might say that Mr Ford never questioned my

engineering recommendations for world trade. Nor, for that matter, did he ever question my commercial and financial recommendations. I never had to write him letters nor argue with him and waste precious time such as we have been wasting. . . .

'If you will not accept my engineering recommendations then I will resign. . . .

'*I would emphasise that the measure of my sincerity and anxiety over this kind of thing are these letters I am sending you, my wakeful nights and the very big offer I made you of not receiving my dividends for a year, or even two years. . . .* My offer was also the measure of my friendship to all you nice people in Massey-Harris-Ferguson. I do hope that nothing will ever happen to hurt our friendship whether or not I should feel compelled to resign.

MY SUGGESTED RESIGNATION

'You know of the high esteem in which I hold you all. You also know that I have no personal ambition *whatsoever* for being anything in the company other than what I am. The truth is I would like to delegate more of my responsibilities.

'All I want to do is to serve the farmers of the world, and serve all of you, with the utmost efficiency possible and with the utmost good cheer and friendship.

'If we cannot run our company in the most friendly, efficient way, and if you think that I am a stumbling block, then will you please suggest some way of disposing of my shares at a fair price and I will resign.'

Shortly after receiving this letter Duncan came to England to try to patch up the situation. Ferguson's personal shareholding in the company was so large that had he decided to dispose of it willy-nilly, and it had fallen into unfriendly hands, the result could have been of grave consequence. Despite the wide divergence of views between the two men, it is interesting that there was no personal acrimony when they met again; in fact they were surprisingly friendly, though their differences of opinion were by then even more marked.

Ferguson let it be understood that many of the best engineers in Coventry would resign with him if the TE60 was not put into production; and since engineering talent had been one of the main assests that Massey-Harris had been out to acquire when they purchased the Ferguson companies, this would have been a serious loss. It was clear, during those meetings with Duncan in April 1954, that the responsibilities that Ferguson assumed he had far exceeded those granted him by the terms of the agreement made when he had sold his company a few months earlier; the right to approve or disapprove engineering design had in his mind become the right to manage engineering activity generally; he even referred to the engineering staff as 'his'. He also had a great deal to say about the company's sales policies, competition from Ford, and the agreement with the Standard Motor Company. Duncan felt that even if Ferguson had sold his business he wanted to continue to manage it.

It was an impossible situation, made even more so when shortly after Duncan's visit Ferguson wrote a twenty-page letter to the Board of Directors of Massey-Harris-Ferguson. That letter began: 'I think you know that I conduct all business in a spirit of fun and good cheer. I am, therefore, sorry to have to write you a letter in very serious vein.' Towards the end, Ferguson wrote: 'If the majority of you gentlemen support Mr Duncan in what he is doing, as against my recommendations, then be assured of my continued friendship and good feeling but I shall want to sell all my stock in Massey-Harris-Ferguson and make an arrangement with you whereby I shall give up all contact with you. I hasten to say that there has been no break between Mr Duncan and myself. All our discussions were on a friendly basis.'

(In passing, it is interesting to note that one of Ferguson's concerns expressed in the letter was not only that the financial strength of Ford might be applied to the tractor market, but also that there would be encroachment of other United States giants, such as General Motors and Chrysler, into the British automobile industry if the British and Canadians did not inject some vigour into their work. Today's situation in the British motor industry shows that his concern was well-founded.)

There was no solution but to accept Ferguson's resignation, buy out his shareholding and obtain the revocation of his veto power on engineering developments. A group of high company officials, led by Duncan and McDougald, arrived in England to bargain with Ferguson. It fell to Duncan to tell him, at the opening of the first meeting, that they had come to accept his resignation.

'Oh I'm not resigning,' Ferguson said jauntily.

'But I have copies of your letters in which you said you would,' replied Duncan.

'Oh you shouldn't take those seriously. I didn't mean it. It was only a threat to bring some pressure to bear.'

'I'm sorry, Harry, but we do take it seriously. We've come to get your resignation,' Duncan told him bluntly.

Ferguson took the news well and seemed quite willing to resign and sell his shareholding. McDougald and Duncan believed their problems to be over and returned to London to settle the details. But the problems were merely beginning; for Ferguson had no intention of giving up without screwing the last possible advantage out of the company, and he seemed to realise that he was in a position of strength. The apparent accord of the first meeting was skin deep and thereafter Ferguson was impossible to pin down; there was even disagreement on the points that had been agreed at the first meeting. It suited Ferguson to be inconsistent and over a period of about six weeks he drove Duncan and McDougald to distraction. They met in London and they met at Abbotswood; they talked incessantly, but only in circles; they fulminated and feuded inconsequentially.

'Jimmy,' said Ferguson at one meeting, 'could you demonstrate one of my tractors?'

'No,' replied Duncan openly.

'Could you repair one if it went wrong?'

'No.'

'And I thought you were an engineer,' Ferguson said scornfully. 'If I'd known you weren't I would never have sold you my company.'

Ferguson wanted to win two specific points before capitulating: he wanted to be able to go back into the tractor industry immediately, in his own name, and he wanted Massey-Harris-Ferguson to provide him with an engineering staff to work in his vehicle development. With regard to the first point, McDougald was very cautious. Ferguson was a name that bore the mark of magic in the world of farming: the right to its exclusive use was important to Massey-Harris-Ferguson. Cannily, McDougald obtained a copy of

Ferguson's birth certificate before pronouncing judgement on Ferguson's demand. When he discovered that 'the old man' was in his seventieth year he reckoned that a five-year embargo on his re-entering the tractor industry would effectively prevent him from becoming a force to be reckoned with again. Ferguson apparently did not think so, for he agreed to the five-year period.

When Ferguson asked for an engineering staff for his vehicle development, to be paid for by Massey-Harris-Ferguson, McDougald was initially amenable.

'We'd be glad to do that, Harry,' he said, thinking that the salaries of a couple of engineers and draughtsmen would be a small additional price to pay to be rid of Ferguson.

'How many are you prepared to give me?' asked Ferguson sharply.

'I don't know anything about your vehicle development. How many do you need?' McDougald said.

'Well, when Henry Ford and I were working together in 1939 I had five hundred engineers, but I'll need far more this time because this is a bigger project.' Ferguson managed to drop this request into the arena with all the calm and innocence of a man asking to borrow a pencil for a moment. But McDougald was a match for him.

'We can forget it then, Harry. I had in mind two,' he said evenly —and won a point because Ferguson was at once enraged.

After five weeks of sporadic wrangling, McDougald gained the impression that Ferguson wanted to raise some cash; it was fairly open knowledge that an Italian inventor, Count Giri de Teramala, was trying to sell Harry Ferguson Research Ltd. the rights of a torque converter transmission he had developed; and the price would probably run to many hundreds of thousands of pounds. In addition, McDougald learnt that Ferguson was negotiating for a Constable painting; and this was not likely to be cheap either. Strange to relate, it was the Constable that finally resolved the impasse between Ferguson and the Canadian negotiators.

McDougald, a sportsman, connoisseur, and collector of many things including veteran cars, was friendly with Oscar Johnson, senior partner in a firm of London art dealers. Filling in time between his inconclusive wranglings with Ferguson, McDougald went to see Johnson about the possibility of having Sir Alfred

Munnings paint a portrait of one of his horses, and he asked Johnson to arrange for them to visit Munnings on the following Thursday afternoon.

'I'm sorry,' Johnson said, 'but I can't manage Thursday because I have a very important negotiation with Mr Ferguson who very much wants to buy a Constable that Lord Glenconner owns.'

The appointment with Munnings was made for another day, but Johnson's words had set McDougald thinking. For some time he had been wondering whether Ferguson really wanted to resign and sell his shares or not; from his contradictory statements over the weeks it was difficult to deduce what really was in his mind; and McDougald was reeling from the comings and goings and the changes of stance. He decided to call Ferguson's bluff. He telephoned John Turner who had been present at most of the meetings.

'John,' he said, 'I'm nearly out of my mind. I've got all sorts of stuff piling up in Canada and I'm sick of Ferguson pushing me over and over and over. You call him and tell him if he's interested in the deal I'm prepared to go to Stow-on-the-Wold next Thursday. I'll sit down and try to make a settlement with him and I'll pay him cash. If he isn't interested tell him its okay by me and I'm leaving for North America on Friday. Just get that message to him.'

It was a clever manœuvre by McDougald, and it bore fruit. The next day McDougald's wife called to see Oscar Johnson and he said:

'I'm so glad you dropped by, I was about to call Mr McDougald and tell him that Ferguson has cancelled his Thursday appointment with me because he has to see some very important gentlemen from North America. So would you please tell Mr McDougald that we can go to see Munnings after all.'

Mrs McDougald, who did not know about her husband's intrigue, returned to their rooms at Claridges and delivered the message.

'That's all I wanted to know!' McDougald said with elation.

For if Ferguson, who loved beautiful things, was prepared to postpone his discussions about the Constable to meet and discuss business instead, he did indeed want to reach a settlement. And sure enough, later that evening, Turner telephoned McDougald.

'Mr Ferguson has asked me to tell you that he has got some other things on, but he'll be glad to see you on Thursday at Abbotswood at ten.'

Duncan and McDougald arrived at the appointed time and once in the library the latter bluntly set the tone for the talks.

'This is the last round, Harry,' he said. 'We've got to reach an agreement today. You are wrecking this company and you are wrecking yourself in the process. I'm prepared to stay until tomorrow morning. If we don't make a deal today I'm going to apply to the courts to have a committee formed to look after your affairs for you.'

Ferguson was much angered by this approach, but he quickly settled down to some serious discussion. In many ways he was extremely gracious, but he played an interesting trick whenever he felt he was being submitted to too much pressure: he ordered the butler to pass the cigars around. Of course they circulated in the silver cigar humidor, which less than a year before the admiring directors of Massey-Harris had presented to Ferguson. Seeing the complimentary inscription they had composed and their own signatures on the box, and comparing the effusively warm atmosphere of a year earlier to the unpleasantness to which they were now subjecting Ferguson, was salutary and slightly shaming.

The negotiators lapsed into double talk again, until about six o'clock in the evening when Ferguson suddenly said:

'Gentlemen, we've all had a long day and I'm tired. Betty and Tony will take you to dinner at Burford and you can come back afterwards and we'll talk some more.'

McDougald had a feeling that agreement would be reached that night and he wanted to be sure that cash would be available in England to meet the cheque he would make out. He therefore decided to allow the rest of the party to go ahead to Burford while he telephoned the Toronto bank that was to advance the large sum of money, approximately $15 million, to pay for Ferguson's shares. (They were to be bought by a holding company in which McDougald was a partner.)

Unbeknown to Ferguson, McDougald went into Maureen

Ferguson's charming little sitting-room, the Boudoir, to place the call to Canada. About twenty minutes later, assured that the money would be cabled to London at once, he was about to come back into the hall when he heard Ferguson ranting at the butler.

'Where is everybody?' he was demanding crossly.

'They've all gone to dinner at the Bay Tree, Mr Ferguson.'

'Well, they didn't show much consideration for me. They were insisting on closing the negotiations today. I'm going to bed and the talks are off until tomorrow. You can tell them that.'

'Harry, have you gone off your head?' McDougald said, emerging into the hall to find Ferguson standing on the stairs. 'You sent the whole damn lot of them off for dinner. You thought I'd gone too, but I was on the telephone and I heard what you said. I'm going over to bring them back and we'll settle this bloody deal tonight.'

'Well, you're an inconsiderate lot,' Ferguson said, and began to harangue McDougald. But McDougald did not stop to listen. He grabbed Ferguson's secretary, Trevor Garbett, and told him to drive to Burford as fast as possible.

'Your dinner is off,' said McDougald rushing into the dining-room of the Bay Tree, just as the party was beginning its soup. 'The whole lot of you get back in your cars. We're heading for Abbotswood as fast as we can.'

Back in the library at Abbotswood, Ferguson began for the first time to talk as though he wanted to close the deal. He passed frequent barbed comments about Duncan, to his face, but by ten-thirty in the evening he had signed his resignation from the chairmanship of the company, relinquished his right of veto on engineering developments, and agreed to accept market price for his shareholding.

The parting at the portals of Abbotswood was quite genial when one considers the events that led up to it. But Ferguson could not resist one last shot at Duncan. As he very courteously shook McDougald's hand he looked at him earnestly. 'Now, Bud,' he said, 'you're a young fellow but obviously you've got access to a lot of money. A fool and his money are soon parted and I'd like to give you a little word of advice.' He pointed to Duncan. 'If you keep on backing this fellow with large sums of money for very long you're going to be on the street.'

It is interesting to reflect for a moment on those protracted negotiations. Without exception, those who took part in them continued to like Ferguson, even if he drove them to distraction from time to time. His explosions were short-lived, slightly theatrical, and were followed quickly by a return to his normal charm and graciousness as a host. It is easy to conclude, because of his mercurial and often irrational mode of negotiating, that the basic premise of his arguments was fallacious. No doubt some of it was, and the TE60 tractor might not have been the blazing success he foresaw. Nevertheless, it is significant that Massey-Harris-Ferguson was being chaotically managed at that time; many mistakes were made, and Duncan was called upon to resign in 1955. It was only the major reorganisation that began in 1956, the merging of the product lines and their marketing under the unified name of Massey-Ferguson, that finally enabled the full potential promised by the merger to be realised. Ferguson may have been over-vociferous in his criticism of Duncan and of his way of running the company, but he had more than a little justification.

Today, as a result of that merger, the company is a world leader in farm machinery. Its president, Albert Thornborough, is an ex-Ferguson man; he was responsible for the vital procurement of material and components when Harry Ferguson Incorporated established its own tractor assembly plant in Detroit after the break with Ford in 1947. His brilliance helped to put the Ferguson company back on its feet then; it was also instrumental in untangling the hideous mess that followed the merger and in establishing the great corporation that is now Massey-Ferguson.

30 *A Fighting Decline*

'Mr Harry Ferguson has made ears of corn grow where before there were none at all. Now he turns from his tractors to a new project—a car that will give unimagined service to the people. Salute the Fergusons for the benefits they provide. They may grow rich, but they make the world grow richer.'

Thus the *Daily Express* lauded Ferguson in an editorial when he announced his complete retirement from the farm machinery industry in order to devote full attention to vehicle development.

'I thank you, Sir,' Ferguson replied in a letter to the editor. 'That is more encouragement to me in the tremendous tasks that lie ahead than all the millions involved.'

But the jaunty words of Ferguson's acknowledgement belied the true feelings that obsessed him much of the time. True, his old brilliance and energy returned in short bursts and he would go 'tootin' around giving everybody hell', but he was ageing, and ailing too. His long bouts of depression reached such depths of misery that he was given a series of shock treatments in 1953 and 1954. To his delight these helped him considerably, but he was also tortured by a variety of ailments which his doctors could neither diagnose nor cure. His insomnia was more intractable than ever, and pains in his thighs often woke him when he had finally managed to go to sleep. From 1953 onwards his vision and hearing deteriorated further; and at various times he had chronic catarrh, pains in the back and fingers, constipation, gum trouble, frequent headaches, anal pains (to alleviate which he had a special rubber cushion with a hole in it), as well as the bothersome burning sensation in the knees which had plagued him for many years off and on. He went to specialists of all types, often seeing three or four during a two-day visit to London. He found most of the new doctors he visited to be excellent—until such time as they began to disagree with his own diagnosis, when they became suddenly incompetent.

Psychosomatic ailments had preyed on him increasingly after

middle age, but it was saddening to those of his staff who worked closely with him to see him so afflicted, to see his characteristic ebullience succumb more and more frequently to what he apologetically referred to as his 'moods'. If he was a slightly comic figure as he carried (or had carried about for him) his rubber cushion with the hole in it, he was also slightly pathetic. He himself realised, in the face of so much inability to diagnose any organic ailment, that his troubles stemmed from other sources; he was particularly impressed by one specialist who told him that he 'worked with too much tension', and by a magazine article entitled 'Stress—the cause of all disease?'

In January 1954 a holiday was suggested and Ferguson and his wife boarded a Comet for Africa. Like almost every other holiday he had ever taken, it was a disaster. They stepped off the Comet at Livingstone to see the Victoria Falls and found themselves in a steam bath; it was an exceptionally hot period and the spray from the Zambesi as it plummeted over the falls raised the humidity to a high level. There to meet them was the managing director of the company's South African interests, John Kelly; Ferguson had hardly been at the Victoria Falls Hotel for an hour before he was asking Kelly about leaving. The next flight was four days hence, and it was the one they were booked on in any case, so he had no option but to stay. He did cheer up a little when he discovered that trains on the long haul between Capetown and the Copper Belt stopped for long periods at the station close to the hotel. His childhood enthusiasm for locomotives returned and he went out each time one stopped to look it over and have a lengthy conversation with the driver and fireman about steam engine design and performance. These interludes were so important to him that he had Kelly go out frequently to see whether a train was coming.

Once in Johannesburg's Carlton Hotel, his first request was for the house physician in the hope of being told that Africa would kill him if he did not leave at once. While Maureen Ferguson and Kelly were waiting in the sitting-room of the hotel suite, Ferguson was being examined by the physician in the bathroom. After close on an hour, and when his wife was becoming alarmed, Ferguson came bursting back into the sitting-room 'much improved'. He had not been examined at all, but he had found that the doctor was fascinated by the Ferguson saga, as related by its protagonist in a lively monologue. It is interesting to note that by this time Ferguson's deafness was increasing noticeably, though he evidently did not want to show it. A result was that he tended to monopolise conversations, and it seems quite likely that some of the difficulties and frustrations encountered by the Massey-Harris directors when attempting to negotiate with him in the summer of 1954 were caused by his not hearing, or not attempting to hear, everything they had to say.

The Johannesburg doctor, who became friendly with the Ferguson's and was primed by Maureen, refused to agree that Africa would kill Ferguson and persuaded him to go to the fine Marine Hotel at Hermanus near Capetown. Here, too, he immediately wanted a doctor. 'He was rapidly reaching the stage where you wanted to regard him as a naughty boy,' Kelly commented. The Marine Hotel did not suit him either, the continuous wind irritated him, and after cutting short their stay they boarded a ship and returned to Britain.

Despite the numerous difficulties of dealing with him, the need to humour him, look after him, and carry his rubber cushion around, Kelly found that Ferguson's few happy moments more than compensated for the trying times of the rest of those weeks. When his old spark returned he was as compelling as ever; one person, a South African tractor dealer who insisted on seeing him briefly in Johannesburg, was interested in illusionism and this set Ferguson talking about the supernatural, a subject on which he had read widely. The meeting, which was to have been of a few minutes only, went on for over two hours as Ferguson talked.

Back in England, in the early spring of 1954, he was ordered complete rest in a nursing home for a month, and he received more shock treatment. But when he returned to Abbotswood he again took up the reins of his many interests, including the fight with the management of Massey-Harris-Ferguson which McDougald forced to a conclusion in July, as already described. Despite his failing health, he had grandiose plans for his vehicle development, and they were plans of such long range that one would expect only a man in the prime of life to have conceived them. Even before selling his companies to Massey-Harris, he had had ideas of obtaining a controlling interest in a major British car manufacturer in order to create a broad-based company that could manufacture cars and tractors to his designs. The necessary finance would have been raised by making a public offering of shares in Harry Ferguson (Holdings) Ltd., the umbrella company that had been established to group the various Ferguson enterprises as they ramified throughout the world in the post-war years. The Standard Motor Company would logically have been the best company in which to obtain a controlling interest since they were already building the Ferguson tractor; but Sir John Black was too independent of spirit, and there was too much friction between him and Ferguson, for such a plan to materialise. The Rover Motor Company was another possibility; negotiations seemed definitely to be heading towards a conclusion when the Rover management suddenly became cautious of allowing Ferguson a controlling interest and suggested instead that he should acquire a minority holding. Ferguson was incensed by the change of heart. 'Under no circumstances would we fool with Rovers, nor co-operate with them on a deal unless I controlled the company,' he said. His attitude in wanting complete control is understandable if we remember the difficulties that being dependent on others had caused in his commercial life. However, the attitude also overflowed into less major issues. For example, the group responsible for the first British Racing Motor asked him to take over the running of the team. This was at the time when the B.R.M. was a supercharged 16 cylinder marvel of engineering design; its silk-rending sound was the highlight of the opening laps of many races—but of the opening laps only because it almost invariably stopped with mechanical failure after a few miles. Ferguson said he would be pleased to take on the task, but only if he had complete and final authority in everything. This was not possible, so the matter ended there.

Considerable liquid assets, almost £3·75 million ($10·5 million), were available to Ferguson after his sale of shares in Massey-Harris-Ferguson. He paid about £500,000 ($1·4 million) for the rights of Count Teramala's torque converter and then continued his attempts to obtain control of a car manufacturer. Sir William Lyons of Jaguar was interested in coming to an arrangement and he examined the inventions being developed at Harry Ferguson Research. His view was that the Ferguson transmission would intrude seriously into a car's passenger space and he told Ferguson so. 'Piffle,' was the typical retort.

Rootes were yet another company that became interested for a while; but it was subsequently the British Motor Corporation that came nearest to taking the plunge. Impressed by the Ferguson prototype car that had been demonstrated to them, B.M.C. decided to carry out a thorough analysis of the likely production costs and sales. The car was to be a vehicle with a shooting-brake type of body to make it useful on farms and estates, but at the same time it was to be a high performance road vehicle with outstanding traction and safety. It would have the convenience of two-pedal driving thanks to the torque converter transmission.

B.M.C. engineers, production and sales experts were as impressed by the vehicle as had been the corporation's management. They calculated that by manufacturing only 350 of the vehicles a week the price at £700 ($1960) would be competitive. They did pinpoint a few disadvantages inherent in the design, but over-all their report was favourable and they recommended building six prototypes. Since B.M.C. had spare productive capacity at the time (1955)—they had just ceased making the Austin Champ military vehicle—all seemed to augur well.

Then, as so often in such matters, there was a hiatus while those responsible were groping towards a decision; and the fact that Count Teramala was continually making detailed design changes to the torque converter while B.M.C. management were deliberating did not make the decision any easier.

Ferguson began to lose the small amount of patience with which he was endowed. He considered the delay to be the fault of Sir Leonard Lord, the head of B.M.C. at the time, and he told him so in several of his inimitable letters. Without any attempt at diplomacy, he took Lord to task for not leaping at the opportunity of adopting the Ferguson inventions, berating Lord soundly, 'for his own good and the good of the country'. Lord, understandably peeved, finally wrote: 'Personally you are a charming fellow but you do write some rubbish, don't you, and I'm getting tired of reading it.' He also challenged Ferguson, who had hinted that Lord was being negligent in his conduct of B.M.C.'s affairs, to attend the next annual general meeting and expose him.

Not long after this acerbic exchange, Lord visited Abbotswood.

'I'm sorry, Harry,' he said, 'but I'm not going to put your car into production. I may be making the biggest mistake of my life, but there it is.'

The collapse of the B.M.C. negotiation was a cruel disappointment to Ferguson, but he took it graciously, as he took most things in face to face confrontations. Had he done more of his dealing in this way he might have achieved better relationships, for his letter-writing usually led to trouble in the end. One associate of many years remarked that he had never known anyone who negotiated by correspondence with Ferguson who did not ultimately fall foul of him. But there were other reasons for his inability to clinch a deal with a motor manufacturer in the mid-fifties. It must be plainly admitted that in old age he became more difficult to deal with, for he saw every issue as black or white and argued on that basis. He had never been an easy man, possessed as he was by a marked sense of his own infallibility, but the very attributes that led to his earlier successes were a stumbling block in later years when they were totally untempered by flexibility. In his youth and middle age, enthusiasm and the conviction that he was right goaded him to a rare degree of perseverance; and usually he was right, as we have seen. However, having been right so often in his life, made him even less willing to listen to opposing views in his old age. His assertiveness, especially when expressed in letters and without the benefit of his personal charm, was irritating beyond measure to men of the standing of Lyons, Lord or Rootes; and he made so many points in his lengthy missives that the primary ones were often lost among the verbiage.

It should not be forgotten either that manufacturers had good reason to be wary of Ferguson. One of the reasons for his fame was his gargantuan lawsuit with Ford, and in the eyes of others, whatever the rights or wrongs of that issue, his evident dogmatism was hardly indicative of a character endowed with sweet reason. Nor can his altercations with the Massey-Harris directors have remained a secret within automotive circles.

A further weakness in his campaign to have his vehicle developments adopted by the motor industry in the mid-fifties was the content of the elaborate demonstrations he laid on at Abbotswood. Even in his days of tractor demonstrating he had tended to make his demonstrations into stunt performances. These were effective for some audiences, but less so to the really well informed, practical spectator. (Witness the supercilious remark passed by a tweedy farmer sitting on his shooting stick as Ferguson was pirouetting a tractor in the confines of a marquee, both to illustrate its manœuvrability and to shelter the audience from the rain: 'Miraculous, but I don't usually do my ploughing in a tent.') Ferguson's car demonstrations in the parkland surrounding Abbotswood were even more gimmicky: his prototypes climbed up and down the steep banks of the Dikler; he deliberately drove his Rolls-Royce into a boggy patch till it was mired to its axles and then towed it out with a four-wheel drive prototype; he treated his visitors, who included the Duke of Edinburgh on one occasion, to some nerve-wracking rides up, down and across steep slopes, and up and down steps as well. But these demonstrations, although proving beyond all doubt the vehicle's cross-country performance, did little to show the extra safety and ease of driving in more normal conditions. And four-wheel drive purely for cross-country use was, after all, far from new in the fifties.

On balance, however, there can be no doubt that Ferguson's personal quirks were a decisive factor in the failure of his attempts to reach agreement with a manufacturer. Indeed, Lord's reply when asked privately why he had decided against making the Ferguson was significant: 'I didn't want Harry Ferguson chasing me around with a red hot toasting-fork,' he said bluntly.

Despite his continuous planning for the future, Ferguson began to develop ever more sentiment for the past as he grew older. A nostalgia for Ulster also gripped him and he returned there for long holidays, renting Lord Downshire's house on Dundrum Bay on one occasion. Some forty-five years earlier, on the sand beaches of that very bay, he had survived crash after crash in his aeroplane and finally triumphed to win the prize offered by the town of Newcastle. It pleased him to be able to look at those beaches again and relive the experience. On other occasions, he and his wife stayed at the Conway Hotel at Dunmurry just south of Belfast. A railway line ran close to the hotel and he liked to be near the tracks when an express hurtled by. He was so fascinated by trains that he arranged to ride on the footplate of a locomotive during one Irish holiday, though he had to cancel the appointment at the last minute. His greatest joy, however, was to be driven around the lanes of his childhood, telling the chauffeur about the people who had lived in the various houses they passed, and recounting stories about them in minute detail.

On one of his visits to Belfast he was talking to Hugh Reid in the offices of the car selling company that still belonged to him (the company that had given him a livelihood and financed his farm machinery research for so many years) when Reid, in a pause in the conversation, raised a delicate topic.

'Er, I saw Willie Sands last week,' he said hesitantly. 'You know, Mr Ferguson, I think he's suffered a great deal since the lawsuit. Here in Ulster people thought pretty badly of him for testifying against you. My feeling is that he regrets what he did and he'd be very glad if you'd see him.'

Ferguson considered this and then smiled whimsically. 'All right, Hugh, for old times' sake I'll see him.'

An appointment was made for a few days later. There was a moment of awkwardness when Sands and Ferguson were facing each other again. This was inevitable since they had worked together more or less closely for over thirty-five years. And Sands's actions had at the end threatened the ruin of all Ferguson's work and hopes.

'Hello, Willie,' said Ferguson finally, holding out his hand. 'It's grand to see you again.' Sands, who despite his periods of discontent in Ferguson's employ, had dedicated himself totally to the work, wept at his ex-employer's reacceptance of him. Ferguson put his arm round his shoulders. 'It's all right, Willie,' he said. 'Everything is forgiven and forgotten.'

Not long after this reconciliation, Ferguson bought Sands an annuity—it is significant that he needed one—so truly had he forgiven what he had once considered a betrayal. To have heard him on the topic of Sands at the time of the lawsuit would have led one to believe that the enmity would be permanent; but however vituperative or vindictive he could be close to a moment of disagreement, with time he returned to a relationship of sincere cordiality, as if nothing had ever been amiss. This transition was as unexpected as it was charming, and there are others whose recollections of bitterness with Ferguson have faded behind warm memories of a reconciliation they had never thought possible.

Ferguson was generous in his way during his visits to Ulster in the fifties. He had very little sympathy for conventional charities and gave to them meagrely, if he gave at all, in response to the frequent requests he received. But person to person charity in favour of old friends and helpers was a different matter. Above all he liked to give cars or television sets to people, sometimes heedless of the fact that the money spent on the television set might have been more useful to them. The concern he felt for those in need was illustrated when he visited the home of an aged farm labourer who had worked for his father. He found the man and his wife in deep poverty and rushed back to the Conway Arms Hotel where Maureen was.

'Those poor people are living in terrible misery,' he told her. 'They've hardly got any clothes to cover their backs. She's about your size, Maureen; you must have a lot of things here you can manage without. Why don't we give them what you can spare as a start?'

Maureen Ferguson went to the wardrobe and pulled out all the dresses, skirts and cardigans that she was not going to need in the immediate future and had them parcelled for him to take to the impoverished couple. She did not demur for an instant, for this was what her husband wanted even if her reason told her it would be more rational to give them the money to go shopping, or have someone purchase clothes for them.

Ferguson's attitude towards providing his own family with financial help was curious. It will be remembered that he had never had a close understanding with any of his brothers or sisters. By the time of his success, few of these brothers or sisters were still alive. Joe, who had originally taken him into Belfast as an apprentice, was himself a prosperous European car importer in New York and he helped look after another brother. A spinster sister was a secretary in a New York office; towards her Harry Ferguson felt a sense of duty and regularly gave her sums of money, finally reaching £400 ($1120) a year tax free. He also felt it his duty, however, to give this sister the benefit of his experience, and thus when he wrote to her, which was seldom, it was often to proffer some advice on how to spend her money to achieve maximum happiness, or to chide her for apparent ingratitude. In fact she was not ungrateful, but she felt that however useful £400 a year might be, it did not substantially alter the kind of life she was able to lead. And in relation to what she assumed her brother to have, it did not give him the right to meddle in her affairs or to expect her to kiss his feet in gratitude. There was therefore no rapprochement between them as they grew elderly and he was irritated beyond reason because she did not write to say how she found pleasure in the money. 'I may have asked you not to write to us every three months on receipt of the money, but really you should,' he fumed with Irish logic in one letter.

With his brother Joe he did establish a closer relationship as the years passed, even though it was only through correspondence; but this was only so once Joe, who was deeply and aggressively religious, had stopped sending Harry tracts and when their letters dealt with memories of their years together in Belfast.

Returns to Ulster were among the happiest periods of Ferguson's latter years. He felt an affinity for the country and its people and for its landscape that none of his long periods in the United States or England had erased. He walked and re-walked the hedgerows of his youth: rain or fine, usually the former. He told the driver of the car to let him out, drive a precise distance along the road to a certain gate and wait for him there. Then stick in hand, raincoat collar turned up and cap pulled down, he set out nimbly across the fields. His driver worried that he might fall into a ditch or boggy patch, but he always appeared, sprightly and with colour in his normally pallid cheeks, at the appointed gateway. The winds and rains of his native lands seemed to erode the stress lines from his face; a youthful buoyancy invigorated his slight form once more, and he was happy.

It is surprising, in view of his pleasure in being in Ulster, that he ever decided to holiday in tropical climes again. But in January 1957, after the usual copious correspondence on every detail including the quality of the coffee, he and his wife went to Jamaica, to Sunset Lodge, a famous hotel with a series of bungalows in the grounds. By coincidence, James Duncan was a part-owner of the hotel, and at the time of the Fergusons' arrival he was occupying the cottage next to theirs. Duncan fully expected either a reopening of the withering hostilities that had marred the year after Ferguson's merging of his company with Massey-Harris, or at least a glacial indifference to his very presence. Instead, and to his total astonishment, Ferguson greeted him as an old friend. It was a return to the same cordial relationship of the pre- and immediately post-merger days of 1953; the bitterness of 1954 had vaporised. Perhaps Ferguson felt sorry for Duncan over his recent forced resignation from Massey-Ferguson, but that would hardly explain the almost incredible offer that he made him: effervescing with enthusiasm over the great automotive advances that Harry Ferguson Research Ltd. were perfecting, he suggested that Duncan become a partner with him and help in their world-wide promotion. Who can know for certain what passed through Ferguson's complex mind? He can hardly have forgotten his differences of opinion with Duncan and his doubts as to Duncan's acumen in running Massey-Ferguson. Had he in the meantime come to think that some of his own opinions had been ill-founded and that he had been unduly harsh in his judgement of Duncan? This would seem to be out of character, for he was not often ready to admit mistakes. The likeliest explanation seems to be simply that he had always been genuinely fond of Duncan, who indeed was a charming and cultivated man. On meeting him again he was captivated by his personality and overlooked the real or imagined defects that had so irritated him during one period of their relationship.

Duncan naturally declined Ferguson's offer but they parted on excellent terms. Most of the Fergusons' pleasure in their holiday departed with him, however. They did not enter into the social life of the hotel, even asking for dinner at their customary, but for the Caribbean uncustomary, hour of 7.00. Thus they sat in solitary state in the large dining-room; and they were in bed by the time most of the guests were beginning to enjoy the evening's social activities. The heat bothered Ferguson and he 'shook his fist at the sun every morning'. The only amusement he seemed to have was in talking with the Jamaicans whose riotous sense of humour engaged him. In the hotel car park one day a Jamaican was holding a soap box meeting; Ferguson walked over and intervened.

'Are you for the Government or against it?' he demanded.

'For it,' came the shout.

'Well you're a damned poor lot of Irishmen,' said Ferguson.

His audience probably did not understand the allusion to the traditional contrariness of the Irish towards authority, but they laughed anyway.

'Are you in favour of Federation?' he demanded. Again he was met with a chorus of assent.

'Well, I'm to be the first president and graft starts right now—hand out!'

The Jamaicans shouted with laughter, doubling over and smacking their thighs in delight. 'No! No!' shouted one of them. 'You hand out to buy our votes.'

One night during which the heat and humidity increased his chronic insomnia, he got up at about 2 a.m. and took extra sleeping pills. They were effective—until he was wakened about an hour later by a series of blows on the head. He turned over and found himself looking into a black face cowled in white cloth. The tall and menacing negro figure, that he took for a woman at first, had a pistol in one hand and a torch in the other; the grotesque apparition was pointing the pistol at him and demanding money. Ferguson struggled to his feet and squared up to his adversary; the intruder towered over the diminutive and elderly man.

At seeing Ferguson apparently ready to give battle, the Jamaican backed off and squeezed the trigger of his pistol; the bullet hit a wall and ricocheted through the roof. Ferguson moved to cover his wife's bed. 'Call the hotel for help,' he said to her.

With calm courage Maureen lifted the telephone and announced that there was an armed burglar in the room. The Jamaican panicked at the course events were taking and loosed another shot; Ferguson felt a searing pain in his left leg. Furious at the treatment he was receiving he decided more firmly than ever that the burglar would get no spoils. He leapt for the door and yanked it open, shouting in his most imperious voice, 'Get the hell out of this. You won't get any money from me.' Thoroughly unnerved, the burglar wasted no time in complying with the order.

The second bullet had gone into Ferguson's leg at the front just below the knee and emerged at the rear of the calf, about half way down. Fortunately it hit no bone. The flesh wound mended well and he and his wife were able to leave for England as planned about three weeks later. He was dreading the meeting with the press at London Airport, but in the event his performance as a pugnacious Irishman delighted the journalists. Pointing his walking stick at his left leg he said, 'The bullet went in there just below the kneecap and straight through my leg. That shows I wasn't running away.' When asked about the intruder he replied, 'I told him to go. He did. He didn't get away with a single damned cent. Tell the Chancellor we saved his sterling.'

The burglar, a thirty-three-year-old former hotel employee who had tried to disguise himself as a woman by wrapping a towel around himself, was subsequently caught and tried. In his defence he claimed that he had been 'savagely attacked' by Harry Ferguson. And a number of Ferguson's acquaintances thought humorously that it might have been true.

31 Shafts of Light in a Stormy Sunset

Several people who were close to Ferguson saw the Jamaican shooting incident as the steepening point in the downhill course of his decline. Even if the flesh wound healed, a psychological aftermath remained that left him suffering from increased depression and made him more unpredictable than usual. At Abbotswood he settled into his inflexible routine, rising punctually, eating punctually, going for his walk over the estate at exactly the same time each day, resting for the same period each afternoon and going to bed at the same early hour each night. (I have often wondered whether if he had led an active social life and gone to bed later his insomnia might not have been improved—a layman's profane judgement no doubt.) The routine was only broken by the occasional demonstration of automobile prototypes, by journeys to Coventry to visit the premises of his research organisation, or by people coming to Abbotswood to see him. Some visitors found him strange in the extreme. One was a relative of McGregor Greer who, at the age of seventy-four, decided she wanted to take a course at the Massey-Ferguson School of Farm Mechanisation in order to improve her performance in ploughing matches. (I had the singular honour of being designated as her instructor.) During her week at Coventry she telephoned Abbotswood for an appointment, and, mindful of Ferguson's views on punctuality, arrived fractionally early. Ferguson was moodily driving a tractor and roller over the parkland, but at exactly the appointed time brought it to a stop and walked up to the house.

'Hello,' he said, 'it's grand to see you again, but I can only spare you four minutes. What shall we talk about?' The friendly tone of voice belied the abruptness of the words, but even so the visitor was nonplussed and unable to suggest a topic of conversation at such short notice. Ferguson was not at a loss, however, and talked earnestly, without pause, about his car development work and the great advances it would bring. The four minutes over, he expressed his pleasure at having seen his visitor and returned to the tractor.

In seeking to explain such behaviour it is easy to conclude that he had not wanted to see the visitor at all, but in fact he was probably engrossed in thought as he trundled around the parkland on the tractor, and therefore time was even more precious to him than usual. His desire to deliver a monologue was probably explained by the deafness that he was so loth to admit to. He cannot have failed to realise that such a performance led to an enhancement of his reputation as a crank; but I suspect that on the basis that unto genius all things are permitted, he rather enjoyed such *enfant terrible* displays.

But Ferguson's last years were not all sullen dusk; the clouds were shot with brilliant light and warm glows—a sunset filled sometimes with the noon dazzle of his best years, and at others with the mellow hues of a man ageing and at peace with the past.

The brilliance showed when, incredible as it may seem, in his seventy-second year he began to plan a come-back into the tractor industry. The idea was born not long after Massey-Harris-Ferguson's introduction in Britain of the new FE 35 tractor model in 1956. This tractor, for which Hermann Klemm was mainly responsible, had been launched about a year earlier in North America, but only after endless battles with Harry Ferguson in the months following the merger. (It will be remembered that it was the tractor that the Detroit engineering staff were putting forward as an alternative to the TE 60, or LTX.) The Ferguson 35 was a very fine tractor, even if it suffered from some grave teething troubles. It did, however, go against several of Ferguson's dogmas: it was more sophisticated than had been the TE 20, the famous 'Fergie' of the war and post-war years, and this meant that it was both more expensive and more complicated to use. To give just one example, it had two hydraulic control levers instead of the single one of the TE 20. Ferguson considered such sophistication a retrograde step because it tended to put the tractor out of the financial and technological grasp of the small farmer, and of the struggling peasant in the developing world.

Massey-Harris-Ferguson considered that it would add to the prestige of the Ferguson 35 if the man whose name it bore would use one on his estate, among his fleet of TE 20s. Bob Annat, who it will be remembered was one of Ferguson's old employees, was therefore despatched to Abbotswood with an FE 35 in an attempt to persuade Ferguson to accept it for his own use.

One of Ferguson's first actions when examining an unfamiliar tractor was to tweak the steering wheel to see how much free movement there was in the steering system. The FE 35 had a recirculating ball system which, though much more pleasant and comfortable to use than that on the TE 20, did have some free play.

'I don't think much of that!' sniffed Ferguson.

He went all around the tractor, peering and poking critically at everything, and then drove it with the cocked ear and minute attention of a test pilot completing his first take-off in a prototype.

'Well,' he said finally, 'apart from the steering, the transmission, the hydraulics and the appearance, I don't suppose it's a bad tractor.'

'Its bonnet design is an advantage over the TE 20,' said Annat. 'It doesn't snag branches and thorns when ploughing near a hedge and whip them back in the driver's face.'

'We could easily change the TE 20 bonnet so it doesn't do that,' replied Ferguson.

That comment set the two men sketching and discussing. Later, over tea in the library, Ferguson suddenly said:

'How much power would we need for a new small tractor to pull two furrows?'

'A minumum of twenty horse.'

'What would it weigh, do you think?'

'Well, the Huddersfield tractor weighed fourteen and a half hundredweight. One should aim for as little over half a ton as possible.'

With Ferguson's enthusiasm rising by the minute, they roughed out the specifications for a tractor that would meet the small farmer's needs.

'That's it, my boy,' he said finally, 'we'll be back in business again.'

The next day he rang up John Chambers and several other members of his old team in an attempt to reconstitute it; not unnaturally he met with a reluctance to drop current activities and go to work with a seventy-two-year-old man on a long-range project without some guarantees. Chambers said politely but firmly that he would need an offer of a contract before he could seriously consider the matter. At first, in his mood of enthusiasm, Ferguson took this reaction as denoting lack of faith in him, and he was hurt. Subsequently he was better able to understand the hesitation, and although none of the long-term employees such as Chambers, Annat or Patterson went back to him, some of the engineers recruited to Coventry after the war did leave Massey-Ferguson and go to Harry Ferguson Research to work on tractor development.

One of Ferguson's immediate problems when he considered returning to tractor development was that he had signed an agreement not to do so for five years after selling his interests to Massey-Harris-Ferguson. This company—which under its new name of Massey-Ferguson was striving to reorganise and amalgamate its product lines—was not likely to turn a blind eye to any infringement of the five-year embargo: the Ferguson name held too much magic in farming circles.

The only solution was to confront the situation and offer his tractor developments to Massey-Ferguson. On the face of it, one would have expected the Massey-Ferguson management politely to decline any further contact with the man who had caused them so much difficulty for almost a year—even if he had sold them a famous name and a brilliant product line. But instead, after a number of contacts, Phillips, McDougald, Thornborough and Young repaired to Abbotswood for a meeting in June 1958. (Young was an ex-Ferguson director, an Englishman who did much to help integrate the companies after the merger.) At that time, Massey-Ferguson was in considerable difficulty. The agricultural machinery market was not flourishing; there were great organisational difficulties within the company; Standard Motor Company's manufacturing prices for the 35 tractor and its newly introduced big brother the 65 were too high; the opposition, and notably Ford, was able to use Ferguson System features as patents expired; and last but not least, the Massey-Ferguson 35 tractor was entering adolescence without ever having recovered from its infant teething problems. Perhaps with these difficulties facing them, the Massey-Ferguson directors thought that the 'old man' might have something worthwhile to offer; if not, at least the courtesy of calling on him could be granted without too much hardship. But Ferguson's offer was concrete: he would advise on improving the 35 tractor in order to simplify it, make it look and perform better, and cost less. Ideally it would be called the Ferguson 35 again, dropping the Massey that had been added to its name. This improvement task would be tied in with a larger range and top secret development programme. One objective of this programme would be to apply the Ferguson (ex-Teramala) transmission to the 35 tractor because 'this new transmission is what the world is waiting for in tractor design'; the other objective would be the design of a 'new small light tractor embodying all the latest Ferguson inventions'. This would be the tractor of the future. A larger model could be designed later on the same principle. These two models would cover such a vast field of tractor sales throughout the world that it would not be worth thinking of anything larger.

'Under no circumstances, nor for any monetary considerations whatsoever, could the rights of this small tractor be *bought* from Ferguson. The future of tractors made on this new principle is too vast to be considered in the light of making a sale.'

It was the Ferguson of twenty years before talking, the man of broad concepts second in importance to none, the man expressing himself with such confidence, enthusiasm and earnestness all rolled together that there was little choice but to believe him or dismiss him as a crank. The Massey-Ferguson directors did however find a middle path: they stalled. Had it really been the Ferguson of twenty years earlier they were dealing with he would have harassed them continually for a decision. Yet the letter he wrote to Phillips on June 25th, 1958, seven months after originally making contact on the issue, was restrained:

'. . . This long delay is certainly bad for all of us. If I read your letter . . . correctly you are suggesting that we wait until September before we come to an arrangement. I am very, very sorry, good friend, but I cannot promise to wait until that time. Other tractor manufacturers are wanting to discuss things with me.'

This uncharacteristic restraint was not only the result of the slowing of tempo of a man moving into old age: Ferguson was occupied on other fronts, for Standard Motor Company seemed to be on the point of adopting the Ferguson torque converter for their cars. This involved much negotiation and time-consuming hoping and planning. And he was also fighting a lawsuit, albeit a very minor skirmish compared to the battle with Ford: Charlie Sorenson had written a book entitled *My Forty Years with Ford* in which, when describing the Ford/Ferguson relationship, he described Ferguson as being 'limited in ability to design'; he added that 'he was not an engineer' and that Ford's reliance on him had been misplaced. More serious still was the insinuation that Ferguson's role in the development of the unit principle and the whole Ferguson System had been of peripheral importance and that Sorenson's was central. The book was withdrawn from British bookshops, and Ferguson obtained an apology and was awarded costs in the High Court.

Perhaps most interesting, however, is the letter Ferguson wrote to Sorenson shortly after the publication of *My Forty Years with Ford*:

> YOUR NEW BOOK
>
> I wish I could congratulate you as warmly on this as I could on other things you have so splendidly done in your career. . . . The story you have told of the relations between Mr Ford, the Ford Company and me, is, unfortunately, a mess of misrepresentations and untruths.
>
> The very wrongful impression you try to convey to the public is that *you* were the man who suggested or created the many inventions in the Ferguson System. . . .
>
> What is the real truth of the matter? Surely this can be proved by the quite long list of patents attached to the tractor. I do not remember that your name figured in any of those inventions. The truth is that at no time in your life did you make one suggestion that was incorporated in the principle of the Ferguson System as marketed in the U.S.A.

Having quoted what he considered misrepresentations in the book, Ferguson went on:

> THE REAL TRUTH
>
> Since you have not told it then someone should. I will now do so to the best of my recollection.
>
> Everybody knows that it was Mr Ford's greatest ambition to do something for the farmer and that he was disappointed in the

original Fordson tractor. When he heard what I had to say about the real needs of agriculture, and saw a demonstration of the Ferguson tractor, he generously said it was his life's dream come true and he would put everything he had behind it, if we could make a deal.

From the day he told you this no man could have been more generous, nor more brilliant, in his efforts to do a good job of manufacture than you were. To this day I take off my hat to you for the energy and genius you put into getting the new tractor into production in record time. You did a magnificent job for the agriculture of your country, and for your Company. Nothing you could ever say, or do, should dim the full credit to which you are entitled.

All this had to do with *manufacture*, and nothing whatever to do with the principle or detail of design, which was not your job.

In addition to doing such a splendid job, your courtesy and kindness to my family and to me, and to all our staff, could not have been excelled. We are still deeply grateful and appreciative to this day.

Those years I spent with Mr Ford and you were the happiest business years of my life. I could not say more than that. As for Mr Ford, I have never ceased to speak in the highest possible terms of his generosity and kindness, and the goodwill and the good humour and cheeriness with which he so splendidly supported and backed the Ferguson System in the interests of world agriculture and his Company.

I cannot now think of one thing you did that would have contributed to the sad and unfortunate breach that arose between Ford and Ferguson. I now believe the cause of that breach was Harry Bennet, coupled with a breakdown in health I suffered under the strain caused by his actions. If you knew all that Bennet did after Mr Ford was taken ill, you would understand the cause of all that happened.

Why you should dim your splendid record by your altogether misleading and untruthful story passes all my comprehension, but there it is.

In conclusion, Charlie, all I will say is that even if you go on blackguarding me for the rest of your life you will not succeed in dimming my appreciation of your kindly friendship and the brilliant job you did during the years we so successfully and pleasantly worked together.

Yours regretfully,
Harry Ferguson

The genuine sentiment expressed in this letter to a man who had insulted him gravely perhaps does much to highlight the fundamental warmth that ran through Ferguson but which was suppressed in so many of his actions and attitudes. With growing age this vein of warmth and sentiment ran to the surface more often. Stronger than ever was his feeling for Ireland. A member of his Abbotswood staff returning to collect some papers one Saturday afternoon heard gay Irish music from the main part of the house, and, with curiosity aroused, he went to investigate: he found Ferguson alone, jacket off and thumbs hitched in waistcoat executing a lively jig. However, the sentiment he exhibited during his last visits to Ireland were less healthy and amusing. For he became obsessed with the greatness of Sir Edward Carson, the brilliant but demagogic Ulster leader mentioned earlier in this book. On one visit to Belfast he insisted on going to Stormont, the seat of Ulster parliament, where there is a large statue of Carson. It was raining, a slanting bitter drizzle, in which Ferguson walked round and round the statue, gazing up at it with a rapt expression. Despite entreaties from Maureen and his sister-in-law, he did not want to leave; and before he finally did he kissed the foot of the statue. Later he wrote to the Dean of Belfast Cathedral to inform him that he was trying to re-awake interest in Carson. He hoped to have a portrait of him painted and present a hundred-guinea (about $295) prize to a cause that would have been dear to Carson. He wanted also to have a silver container made into which a sod from Abbotswood would be placed and the whole set on Carson's tomb in Belfast Cathedral. A tribute to Carson would be engraved on a plaque and would finish with the verse that Mark Twain wrote for his daughter's grave:

Warm summer sun
Shine kindly here
Soft southern winds
Blow gently here

Ferguson was profoundly moved by this and similar sentimental verse. At Abbotswood he often recited long passages from Kipling to members of his staff. 'Isn't that beautiful?' he would say at the end. He liked to be read to and would listen with repeated pleasure to *The Cloister and the Hearth*, the essays of Ingersoll, Conan Doyle's *White Company*, Jack London's *Call of the Wild*, *The Rubaiyat*, and Darwin's *Origin of Species*.

In the depressions which again swept over him in 1960 he sometimes passed many days in which he hardly moved from his library, apart from his excursions into the garden or park. Alone, consumed by bleak thoughts, he lay inert on the library sofa, inaccessible to those who cared for him and wanted to help, locked in a prison of dejection for which no one had the key. To see a man often so buoyant dragged into such pits of misery was saddening. He was more than ever obsessed with his own health, or rather lack of it, and was constantly in touch with his local doctor and numerous specialists. But they could do little to alleviate his suffering.

However, after such a period, he would quite suddenly revert to his dynamic self. In May 1960 he went, as he often did, to a motor race at Silverstone, the circuit in Northamptonshire where many important events have been held. His companion on such occasions was often his young estate foreman, Peter Warr, an open-faced and friendly man of whom he was fond and whose cheerful undemanding company he enjoyed. At Silverstone Ferguson wandered everywhere, talking to mechanics and drivers and scrutinising the cars in the pits and paddock. He had many acquaintances in the racing fraternity and since he was so well informed on car design his visits and discussions were welcomed, even if they became all engrossing; at one race he was still on the starting grid talking to a driver after the signal to clear the track of non-participants had sounded. In his distinctive corduroy cap and raincoat he was oblivious to surrounding events as he leant into the cockpit in earnest conversation. Only the announcer saying, 'If Mr Ferguson would be good enough to leave the track we'll start the race,' finally moved him.

The most important result of that May 1960 visit to Silverstone was his decision to build a racing car incorporating the various inventions that Ferguson Research had been developing. Many technical innovations have been put to the test of motor racing before being accepted for more general use, and so Ferguson believed that building a car incorporating his team's inventions and placing it in the inferno of Formula 1 competition might lead to the adoption of those inventions by the motor industry. He had often talked about building a racing car, but this time he was more determined and Project 99 was launched.

Like everything else with which he had ever been involved, Project 99 was characterised by the taking of infinite pains. All the drawings for the car were printed on green paper instead of the usual white; and the production facilities at Harry Ferguson Research were instructed that not only were they to give priority to the components shown on green drawings but that there were to be no, absolutely no, concessions with regard to tolerance or finish. Only 100 per cent accuracy in all stages of manufacture, from raw material to finished component, would be acceptable. Perfection would only be just good enough.

By July 1960, Ferguson was convinced that he would soon be able to realise his aim of returning to the tractor business too. Indeed he wrote to A. G. B. Owen to inform him that he would soon be back in the industry 'right up to the crown of my hair'. He had even fixed a target price of £400 ($1120) for the new small Ferguson tractor—against the average price of £650 ($1820) for competitive machines—and £550 ($1540) for a big new tractor, as opposed to about £760 ($2130) for other models of similar size. Also in July 1960, he was negotiating with the Rootes Group again in the hope that this motor manufacturer would help to launch his 'far-reaching plans'.

Lengthy letters were leaving Abbotswood by the score again; he wrote to anyone and everyone about his plans. He sent a two-page letter to the Gillette razor company telling them what was wrong with their blade design, but he did not let slip the chance of informing them that, 'Our new inventions will constitute the most romantic, the most profitable and the greatest industrial development in the history of world commerce. I have been working on this chassis for forty years.'

But this was to be the last of Ferguson's great spells of activity, the last ray of light slanting through the grey clouds of the sunset of his life. In such periods of activity during his earlier years he had been a superman; he had tackled problems against all odds and solved them; he had confronted hostility and opposition to his apparently eccentric ideas and proved them right; he had crusaded for many causes, isolated in the solitude that comes to all original thinkers, and he had converted to his ideas people of much greater inherent intellectual power than his own. But now, at seventy-five years of age, his energies were running out. The bravura that had been the epitome of so much of his life had a hollow ring to it. His reputation for being difficult to work with had not diminished, for he tended to be overbearing in his dealings. As one journalist commented, 'According to Coventry manufacturers, you can't work *with* him, you have to work *under* him. And why should people like Lord and Tedder work under Harry Ferguson?' The press was indeed rather bitter about the non-appearance of the Ferguson car for which hopes had been built so high in the fifties. (It was still not generally known of course that a racing car was being constructed, nor that Renault had adopted the Ferguson/Teramala torque converter.)

By the late summer of 1960, his last burst of energy was exhausted; gradually he sank back into a bout of depression again, and more shock treatment was prescribed. He finished the series of treatments about October 22nd and on the 23rd, a Sunday, his daughter Betty visited Abbotswood to see him. She was distressed to find him inert and suffering from the characteristic loss of memory that usually follows shock treatment. He lay on his bed listless and remote and his daughter returned worried to London where she was living. She telephoned her mother on the Monday and was delighted to hear that her father had left his bedroom and was jauntily taking up his activities again.

The next morning, October 25th, the footman Scott took Ferguson his shaving water at precisely 7.45 as usual and drew back the curtains.

'Good morning, Scott. What kind of a day is it?' Ferguson asked brightly.

'It's not too bad. How are you feeling, Mr Ferguson?' Scott asked, encouraged by his employer's apparent cheerfulness.

'I'm getting better,' Ferguson replied, pulling on his dressing-gown and making for the door. For every day as soon as he had been called he went to his wife's room to wish her good morning with a kiss. His unvarying habit thereafter was to return to his bathroom to shave and bath. In the meantime, the domestic staff prepared breakfast and at 8.30 precisely carried it to Maureen Ferguson's room; her husband joined her there for his fruit, scones and punctiliously prepared coffee, and his arrival seldom missed coinciding with that of the breakfast by more than a minute or so. On that morning of October 25th, 1960, however, the breakfast had arrived in Maureen Ferguson's room and the coffee had been cooling for close on five minutes, yet there was no sign of Ferguson. His wife went to investigate this uncharacteristic tardiness: she found him submerged in his bathwater. The butler and footman came rushing at her frantic summons. Tenderly they lifted the slight body from the bath, but it was too late. Harry Ferguson was dead.

A post-mortem examination was carried out, and on November 8th, 1960, an inquest was held at Cheltenham. The pathologist stated that Ferguson had been in good physical health for his age but that 'from a toxological examination of the organs it appeared that (he) had taken the equivalent of fifty-seven three-grain tablets of Phanodorm and fifty-seven one-grain tablets of sodium amytol'. This amounted to about twice the lethal dose. 'He died from an overdose of barbiturate tablets but there is no evidence to show whether they were self-administered or ingested accidentally,' the coroner said.

During the two-hour hearing Ferguson's own doctor described how his patient had been found unconscious in his bath, in very similar circumstances, in July 1959. On that occasion he had been rushed to Moreton-in-the-Marsh hospital and revived. The doctor also related how Ferguson had always kept a record of his drug consumption and tried to reduce the dosage. Before the jury left for their fifty-minute deliberation they were reminded of the absence of any written or verbal statement by Ferguson indicating an intention to take his own life. They returned an open verdict, and thereby the mystery was officially left unresolved.

Among people who were close to Ferguson there are those who say that he would never have committed suicide; others state that they have no doubt he did. His doctor was never able, despite repeated attempts, to make him talk about the occasion in July of the previous year when he was found unconscious in the bath. On the other hand, most sufferers from depression make an attempt on their life sooner or later unless carefully controlled with the modern drugs that were not generally available in 1960. Maureen Ferguson, who was devastated by her loss, steadfastly maintained that he had not deliberately taken his life. This calm and gracious person who had dedicated herself to help her husband achieve his ambitions suddenly found herself without a goal, without a reason to live. She never fully recovered from the shock, but with great fortitude she struggled on through a serious illness of her own. She died some five years after her stormy-petrel husband.

It is worth recording that Ferguson remained an agnostic to the end. Not for him the change of heart that impending death brings to many. As per his instructions, his body was cremated and the ashes spread from an aircraft over Abbotswood.

In summary what can one say about such a man? The results of his life's work are evident for all to see in the sophisticated farm machinery that in some countries today enables such a small proportion of the population to feed the rest. For Ferguson did more than anyone else before or since to apply real engineering to the production of the food we need to survive. The principles of the Ferguson System opened up new horizons, and even if certain manufacturers today achieve similar results through slightly different designs, the basis remains the one evolved by that punctilious and stubborn Ulsterman. There can be few who would deny the importance of his contribution. Some detractors whom I spoke with when researching this book stated that the Ferguson System, or something very like it, would have come into existence anyway, sooner or later. But surely the same could be said about the electric bulb, the aeroplane, or the internal combustion engine; yet who dares to denigrate the work of Edison, the Wright Brothers or Otto? If one stops to investigate the motivation behind Ferguson's detractors one finds invariably that they criticise his work because they disliked him personally or fell foul of him. And as a man his style was not to everyone's taste: he was too opinionated, self-assured and intolerant to be acceptable to some people; yet others found him 'quite delightful'. As was stated at the beginning, he was a man of contradictions, but his singleness of purpose drove him to success. And we may not have seen the last of his influence, for the Ferguson Formula of all-wheel control may yet, more than ten years after his death, bring great change to the motor industry. P 99, the Formula 1 racing car that he gave the order to build, achieved some noteworthy success in the hands of Stirling Moss and proved the extraordinary roadworthiness that results from the system of four-wheel drive and anti-lock braking. Harry Ferguson Research has continued to promote the inventions and after a long drawn out battle seems at the time of writing to be on the point of gaining their acceptance by the motor industry. From the motoring press they have been receiving outstanding plaudits for years, but the industry has been loth to put into production what amounts to an amazing safety feature because so far the public has shown little interest in buying safety. But things are changing, as they should be when it is remembered that every year in the world about 200,000 people die and 15 million are injured in road accidents.

Even if the name of Ferguson becomes as famous in automobile circles as it is in the tractor industry, it will not change the fact that he was a man in many ways to be pitied. He was essentially gloomy. Success brought him no happiness, only an urge to succeed in other fields. Lonely, despite having people around him, persecuted by his psychosomatic ailments, so obsessed by his objectives that he was without sense of proportion, he was a tragic figure. A self-appointed saviour of mankind needs a sense of humour too; Ferguson's sense of humour was not all-embracing, for it seldom included himself. Without Maureen he would have made little progress; with a greater sense of proportion he might have made even more. The failure of the Ferguson System to bring farm mechanisation to the agriculture of the whole world was a major disillusion to him and one that left him with a sense of defeat in the midst of much victory. Why else would he in his very last years have tried to launch a new and even simpler and cheaper tractor for the world's farmers? Much he did accomplish, and the following illustrates what his inventions meant to some small farmers, even if in the world generally the tractor still cannot be used economically.

In the spring of 1948, the first Ferguson tractor, a TE 20,

arrived in the Adana area of Turkey and the newly-appointed dealer for the area put it in his show window in the town. Curiosity was aroused by the attractive silhouette of the tractor with its neatly mounted plough, but it was considered a toy compared to the large tractors that dominated the local market. Almost no one enquired about its price. Then a Hodja in his robe and turban stopped and gazed intently at the tractor and finally entered the shop to ask whether it could be demonstrated on his farm. In order not to run the tractor unnecessarily it was loaded onto a horse cart and carried the ten miles to the Hodja's field where he asked that it be put to work. The little tractor turned the crumbly soil and raised a cloud of dust as the Hodja walked closely behind. Then, still walking, he began to shake his head. When the dealer stopped and asked if something was the matter, the Hodja merely grunted and asked him to continue. Four lengths of the field he ploughed with the Hodja walking behind shaking his head. When he finally stopped and switched off the tractor, the dust-covered Hodja lent his elbows on the tractor bonnet and cupped his chin in his hands. Astonishment was etched in every line of his face.

'Tell me, Rashid Bey,' he said finally, 'am I dreaming or is this true? Can such a small machine, a toy, do so much work? Let us have coffee and you will tell me about it.'

Over coffee, in the house, the dealer explained the Ferguson System. The Hodja brought out banknotes to the full cash value of the tractor and plough and passed them over in settlement. But to his surprise, on emerging from the house, the dealer saw that the tractor had disappeared. He asked what had happened to it and the Hodja smiled. 'I told my man to take it away and lock it up. I am taking no risks. I want this machine and no other because I do not believe you have another which will do the same.'

That Hodja, and farmers like him all over the world, marvelled at the Ferguson System. For many it was a dream come true. To those in Wales the TE 20 was the only tractor that had ever met their needs; in Welsh the word tractor is 'Fergie'. A man could ask for little better as a tribute, but another that would have pleased Harry Ferguson greatly was the cable that reached Abbotswood from Massey-Ferguson shortly after his death:

THE HOURLY EMPLOYEES OF THE DETROIT TRACTOR PLANT RECEIVED WITH DEEP REGRET THE SAD NEWS OF THE PASSING OF MR HARRY FERGUSON STOP WE CONSIDERED HIM OUR FRIEND WHO SYMPATHISED WITH OUR VIEWS AS WORKING PEOPLE STOP THE MANY FINE THINGS HE DID FOR US WILL NEVER BE FORGOTTEN STOP PLEASE ACCEPT OUR DEEPEST HEARTFELT SYMPATHY

SIGNED HOURLY RATED EMPLOYEES OF UNIT 174 VAW MASSEY-FERGUSON DETROIT.

Index

Index

This is a reprint of the second printing (1973) of the John Murray hardback edition, with minor alterations.

*

Also from Old Pond Publishing

The Ford Tractor Story, Stuart Gibbard (published jointly with Japonica Press)
Ferguson Tractors, Country Films (video).

Old Pond Publishing
104 Valley Road
Ipswich IP1 4PA
United Kingdom

Tel: 01473 210176/254984
Fax: 01473 210176
Email: r.smith@virgin.net